Applied and Numerical Harmonic Analysis

Series Editor

John J. Benedetto
University of Maryland

Editorial Advisory Board

Applied and Numerical Harmonic Analysis

(Continued after index)

Kai Borre
Dennis M. Akos
Nicolaj Bertelsen
Peter Rinder
Søren Holdt Jensen

A Software-Defined GPS and Galileo Receiver

A Single-Frequency Approach

Birkhäuser
Boston • Basel • Berlin

Kai Borre
Danish GPS Center
Aalborg University
Niels Jernes Vej 14
DK-9220 Aalborg Ø
Denmark

Dennis M. Akos
Department of Aerospace Engineering
University of Colorado
Boulder, CO 80309
U.S.A.

Nicolaj Bertelsen
Hornbækvej 116
DK-9270 Klarup
Denmark

Peter Rinder
Vivaldisvej 148
DK-9200 Aalborg SV
Denmark

Søren Holdt Jensen
Department of Electronic Systems
Aalborg University
Fredrik Bajers Vej 7
DK-9220 Aalborg Ø
Denmark

Cover design by Joseph Sherman.

MATLAB® and Simulink® are registered trademarks of The MathWorks, Inc.

Mathematics Subject Classification (2000): 34B27, 49-XX, 49R50, 65Nxx

Library of Congress Control Number: 2006932239

Additional material to this book can be downloaded from http://extras.springer.com

ISBN 978-0-8176-4390-4   ISBN 978-0-8176-4540-3 (eBook)
DOI 10.1007/978-0-8176-4540-3

Printed on acid-free paper.

9 8 7 6 5 4 3 2 1

Applied and Numerical Harmonic Analysis (Cont'd)

J.A. Hogan and J.D. Lakey: *Time-Frequency and Time-Scale Methods* (ISBN 0-8176-4276-5)

C. Heil: *Harmonic Analysis and Applications* (ISBN 0-8176-3778-8)

K. Borre, D.M. Akos, N. Bertelsen, P. Rinder, and S.H. Jensen: *A Software-Defined GPS and Galileo Receiver* (ISBN 0-8176-4390-7)

Contents

Preface

Software-defined radios (SDRs) have been around for more than a decade. The first complete Global Positioning System (GPS) implementation was described by Dennis Akos in 1997. Since then several research groups have presented their contributions. We therefore find it timely to publish an up-to-date text on the subject and at the same time include Galileo, the forthcoming European satellite-based navigation system. Both GPS and Galileo belong to the category of Global Navigation Satellite Systems (GNSS).

SDR is a rapidly evolving technology that is getting enormous recognition and is generating widespread interest in the receiver industry. SDR technology aims at a flexible open-architecture receiver, which helps in building reconfigurable SDRs where dynamic selection of parameters for individual modules is possible. The receiver employs a wideband analog-to-digital (A/D) converter that captures all channels of the software radio node. The receiver then extracts, downconverts, and demodulates the channel waveform using software on a general-purpose processor. The idea is to position a wideband A/D converter as close to an antenna as is convenient, transfer those samples into a programmable element, and apply digital signal processing techniques to obtain the desired result. An SDR is an ideal platform for development, testing of algorithms, and possible integration of other devices. We have chosen MATLAB® (version 7.x) as our coding language because MATLAB is the de facto programming environment at technical universities, it is a flexible language, and it is easy to learn. Additionally, it provides excellent facility for the presentation of graphical results.

The concepts of the present project crystallized nearly 10 years ago. At that time the technology was not mature enough to be fully realized. Today the various elements of this interactive book and companion DVD work, and work together:

- The main text, which describes a GNSS software-defined radio;

- A complete GPS software receiver implemented using MATLAB as well as sample data sets—available on the companion cross-platform DVD—allowing readers to change various parameters and immediately see their effects;

- A supplementary, optional USB GNSS front-end module, which eventually converts analog signals to digital signals on a Windows or Linux system (ordering instructions for this hardware can be found on the bundled DVD). The signals are imported into MATLAB via the USB 2.0 port.

With all three elements, readers may construct their own GNSS receivers and will be able to compute a position. Is this not exciting?

A GNSS software receiver is by no means a simple device. The one presented in this book is a single-frequency receiver using the C/A code on L1 for GPS. This choice has been made for various reasons. First, it keeps the receiver architecture fairly simple. Second, with the orbital, clock and ionospheric corrections, as well as the integrity provided by satellite-based augmentation systems (SBAS) such as the European Geostationary Navigation Overlay System (EGNOS) and Wide Area Augmentation System (WAAS), L1 accuracies rival that of dual frequency GNSS receivers. Only future GNSS wideband signals or additional power will offer performance improvements over the L1 band. Lastly, the forthcoming European Galileo system will offer an L1 component that will provide greater availability with its additional full constellation of satellites. We include a description of the binary offset carrier (BOC) modulation, which is to be used in Galileo, to allow readers to get acquainted with the new signal type.

The book considers the design of the front-end module, explaining how it is possible to split the enormous amount of information contained in the antenna signal into parts related to the individually tracked satellites. Details on purchasing the front-end module designed for this text can be found at: http://ccar.colorado.edu/gnss.

A complete GPS software receiver was implemented in MATLAB. The receiver is able to perform acquisition, code and carrier tracking, navigation bit extraction, navigation data decoding, pseudorange estimation, and position computations. We were told that nowadays a textbook comes with overheads prepared for the classroom so that the lecturer's job is minimized. Actually we have used most of the present text for various courses during 2004 and 2005, and we are happy to make this material readily available; it can be downloaded from: gps.aau.dk/softgps.

What follows here is a brief overview of the book's contents and coverage of topics. Chapter 1 contains a short introduction to deterministic and random signals. All GNSS codes build on an elementary pulse signal. We investigate its amplitude spectrum, autocorrelation function, and its Fourier transform, viz. the power spectrum. The important sampling process is described. Next, systems and especially linear time-invariant (LTI) systems are dealt with. This first chapter should give the mathematically oriented reader a chance to follow the rest of the book. Generally, the book is written by signal processing people for signal processing people.

The signal structure and the navigation data for GPS are described in Chapter 2. We deal with the pseudo random noise (PRN) sequences and their correlation properties in great detail. Doppler frequency shift and code tracking is mentioned. Finally the navigation data and their format are described.

Chapter 3 has a flow similar to the preceding chapter. We concentrate on the L1 OS signal and binary offset carrier (BOC) modulation, which is new in connection with navigation signals from satellites. Coherent adaptive subcarrier modulation (CASM) as well as cyclic redundancy check (CRC), forward error correction (FEC), and block interleaving are described in detail. Finally, the message structure and contents (navigation data) are dealt with.

Chapter 4 gives an introduction to the GNSS front-end module consisting of an antenna, filter, amplifier, mixer, and analog-to-digital converter. The resulting raw sampled data characteristics are described, and finally miniature GNSS front-end designs, in the form of application specific integrated circuits (ASICs), are discussed.

Chapter 5 provides an overview of the high-level signal processing within a GNSS receiver from the most commonly used structure of a GNSS receiver. This starts with a description of the functionality of an individual channel, as well as a brief description of the required signal processing from acquisition all the way through the position solution.

Chapter 6 describes in detail GPS signal acquisition, or identifying the signals present in the collected data sets. The implementation of both serial and parallel search acquisition methods are described.

Chapter 7 describes code and carrier tracking and data demodulation. Various delay lock loops (DLL) discriminators are described. A phase lock loop (PLL) and a frequency lock loop (FLL) are often used to track a carrier wave signal. These elements are critical to refining the precise measurements provided in GNSS. A special investigation is devoted to describing the envelope of multipath errors under various conditions.

Chapter 8 recovers the navigation data and converts them to ephemerides. An ephemeris makes the basis for computing a satellite position. Next we demonstrate how to estimate the raw and fine parts of the transmit time. With these data we introduce a computational model that delivers the receiver position. In order to make the text more complete, we have added material about error sources, time systems, and several relevant coordinate transformations. The concept of dilution of precision (DOP) is studied as well as coordinate and time reference frames. Finally, a positioning procedure for a combined GPS and Galileo receiver is given.

Next follows a final chapter with problems. They are of different complexity; we hope they will stimulate the reader to work with the comprehensive material contained in this book. The ultimate goal of the authors will be achieved if the reader is able to change key parameters (loop noise bandwidth and damping ratio of DLL and/or PLL, correlator spacing, acquisition thresholds, sampling and intermediate frequencies, elevation mask, acquisition bandwidth) and study the effect of those changes.

The book ends with two appendices. Appendix A gives an overall exposition of the MATLAB program. The actual code is well documented, but we still find it useful to add some general comments on the structure. Finally, we enumerate the differences between GPS and Galileo signals in preparation for an eventual software implementation of the Galileo L1 OS signal.

Appendix B contains a description of how to implement a simulation facility for GPS signals. This is done in Simulink. Based on available literature, a Galileo L1 OS signal generator is likewise described in Simulink. The generated spectrum resembles the received Galileo In-Orbit Validation Element A (GIOVE-A) spectrum very well!

The contributions of Darius Plaušinaitis in making a total revision of the code, reworking and creating of numerous figures, testing, coding the Galileo simulator, and writing the USB driver for Windows cannot be underestimated. Henrik Have Lindberg contributed to the clarification of certain aspects of the CASM. Troels Pedersen improved the multipath description. Kristin Larson should also be recognized for her contribution and testing of the MATLAB algorithms. The folks at SiGe Semiconductor, particularly Stuart Strickland and Michael Ball, should be acknowledged for their efforts in helping to develop the available USB GNSS front-end module. Additionally Stephan Esterhuizen and Marcus Junered should be recognized for their Linux USB driver and application development associated with the module.

All the constructive help from several reviewers is highly appreciated.

Substantial financial support from the Danish Technical Research Council for Nicolaj Bertelsen, Darius Plaušinaitis, and Peter Rinder is acknowledged. A grant from Det Obelske Familiefond has supported the development of the USB driver for Windows.

We are grateful that Birkhäuser Boston is our publisher. Today we have marvelous production tools at our disposal like TEX. We know we are not the only ones who appreciate this product. We used LATEX 2_ε and $\mathcal{AMS}$-TEX, and the book is set in the 10/12 fonts Times and MathTime from Y&Y. The very competent page layout and numerous intricate TEX solutions are due to Frank Jensen. John D. Hobby's MetaPost should also be mentioned. Twenty-five figures were created in this environment, which is a mixture of Donald Knuth's METAFONT and PostScript from Adobe. Using it is sheer joy.

Aalborg, August 2006

Kai Borre	*Dennis Akos*	*Nicolaj Bertelsen*
Aalborg University	University of Colorado	Private consultant
borre@gps.aau.dk	dma@colorado.edu	nicolaj@bertelserne.dk

Peter Rinder	*Søren Holdt Jensen*
GateHouse A/S	Aalborg University
pri@gatehouse.dk	shj@kom.aau.dk

List of Figures

List of Tables

Abbreviations

ACF	Autocorrelation Function
ADC	Analog-to-Digital Converter
AGC	Automatic Gain Control
ASIC	Application-Specific Integrated Circuit
AWGN	Additive White Gaussian Noise
BOC	Binary Offset Carrier
BPSK	Binary Phase-Shift Keying
CASM	Coherent Adaptive Subcarrier Modulation
CDMA	Code Division Multiple Access
CRC	Cyclic Redundancy Check
DAB	Digital Acquisition Board
DFT	Discrete Fourier Transform
DLL	Delay Lock Loop
DOP	Dilution of Precision
DVD	Digital Versatile Disc
ECEF	Earth Centered, Earth Fixed
EGNOS	European Geostationary Navigation Overlay System
FEC	Forward Error Correction
FFT	Fast Fourier Transform
FLL	Frequency Lock Loop

GAST   Greenwich Apparent Sidereal Time

GDOP   Geometric Dilution of Precision

GIOVE-A   Galileo In-Orbit Validation Element A

GNSS   Global Navigation Satellite System

GPS   Global Positioning System

GPST   GPS Time

GST   Galileo System Time

HOW   Handover Word. 17-bit truncated version of TOW

IERS   International Earth Rotation and Reference Systems Service

IF   Intermediate Frequency

IOD   Issue of Data

JD   Julian Date

L1   The GPS and Galileo frequency $f_{L1} = 1575.42\,\text{MHz}$

LFSR   Linear Feedback Shift Register

LHCP   Left Hand Circular Polarization

LNA   Low Noise Amplifier

LO   Local Oscillator

LSB   Least Significant Bit

LTI   Linear Time-Invariant

MSB   Most Significant Bit

NAD 27   North American Datum 1927

NCO   Numerically Controlled Oscillator

OS   Open Service

PLL   Phase Lock Loop

PRN   Pseudo Random Noise

PSD   Power Spectral Density

RF   Radio Frequency

RHCP   Right Hand Circular Polarization

RINEX   Receiver Independent Exchange Format

RNSS   Radio Navigation Satellite Service

SAW   Surface Acoustic Wave

SBAS   Satellite-Based Augmentation System

SDR   Software-Defined Radio

SISA   Signal In Space Accuracy

SNR   Signal-to-Noise Ratio

SOW   Second of Week

SPS	Standard Positioning Service. Position computations based on C/A code signals
SVN	Space Vehicle Number
TAI	Temps Atomique International
TLM	Telemetry word. Eight-bit preamble used to synchronize the navigation messages
TOW	Time of Week. The GPS week starts midnight Saturday/Sunday
UHF	Ultra High Frequency
USB	Universal Serial Bus
UT	Universal Time
UTC	Universal Time Coordinated
UTM	Universal Transverse Mercator
VCO	Voltage Controlled Oscillator
VSWR	Voltage Standing Wave Ratio
WAAS	Wide Area Augmentation System
WGS 84	World Geodetic System 1984

1

Signals and Systems

The concepts of a signal and a system are crucial to the topic of this book. We will consider time as well as frequency domain models of the signals. We focus on signals and system components that are important to study software-defined GPS and Galileo receiver design. For a detailed treatment of signal and system theory, we refer to the many standard signal processing textbooks on the market.

1.1 Characterization of Signals

In satellite positioning systems, we encounter two classes of signals referred to as deterministic and random signals. Deterministic signals are modeled by explicit mathematical expressions. The signals $x(t) = 10\cos(100t)$ and $x(t) = 5e^{50t}$ are examples of deterministic signals. A random signal, on the other hand, is a signal about which there is some degree of uncertainty. An example of a random signal is a received GPS signal: the received signal contains beside the information bearing signal also noise from disturbances in the atmosphere and noise from the internal circuitry of the GPS receiver; see Chapter 4 for more details.

Now some basic topics on deterministic and stochastic signal theory are reviewed and simultaneously we establish a notation.

A reader familiar with *random processes* knows concepts like autocorrelation function, power spectral density function (or power spectrum), and bandwidth. These concepts can be applied for deterministic signals as well, and that is exactly what we intend to do in the following. There are several good sources about random processes; see, for example, Strang & Borre (1997), Chapter 16.

1.1.1   Continuous-Time Deterministic Signals

Let us consider a deterministic *continuous-time signal* $x(t)$, real- or complex-valued with finite energy defined as $\mathcal{E} = \int_{-\infty}^{\infty} |x(t)|\, dt$. The symbol $|\cdot|$ denotes the absolute value, or magnitude, of the complex quantity. In the frequency domain this signal is represented by its *Fourier transform*:

$$X(\omega) = \int_{-\infty}^{\infty} x(t)e^{-j\omega t}\, dt, \tag{1.1}$$

where $j = \sqrt{-1}$ and the variable ω denotes angular frequency. By definition $\omega = 2\pi f$ and the units for ω and f are radian and cycle, respectively. In general, the Fourier transform is complex:

$$X(\omega) = \Re\big(X(\omega)\big) + j\Im\big(X(\omega)\big) = \big|X(\omega)\big|e^{j\,\arg(X(\omega))}. \tag{1.2}$$

The quantity $X(\omega)$ is often referred to as the *spectrum* of the signal $x(t)$ because the Fourier transform measures the frequency content, or spectrum, of $x(t)$. Similarly, we refer to $|X(\omega)|$ as the *magnitude spectrum* of $x(t)$, and to $\arg\big(X(\omega)\big) = \arctan\big(\Im(X(\omega))/\Re(X(\omega))\big)$ as the *phase spectrum* of $x(t)$. Moreover, we refer to $|X(\omega)|^2$ as the *energy density spectrum* of $x(t)$ because it represents the distribution of signal energy as a function of frequency. It is denoted $\mathcal{E}_x(\omega) = |X(\omega)|^2$.

The *inverse Fourier transform* $x(t)$ of $X(\omega)$ is

$$x(t) = \frac{1}{2\pi} \int_{-\infty}^{\infty} X(\omega)e^{j\omega t}\, dt. \tag{1.3}$$

We say that $x(t)$ and $X(\omega)$ constitute a *Fourier transform pair*:

$$x(t) \leftrightarrow X(\omega).$$

The energy density spectrum $\mathcal{E}_x(\omega)$ of a deterministic continuous-time signal $x(t)$ can also be found by means of the (time-average) *autocorrelation function* (ACF) of the finite energy signal $x(t)$. Let * denote complex conjugation, and then the ACF of $x(t)$ is defined as

$$r_x(\tau) = \int_{-\infty}^{\infty} x^*(t)x(t+\tau)\, dt, \tag{1.4}$$

and the energy density spectrum $\mathcal{E}_x(\omega)$ of $x(t)$ is defined as

$$\mathcal{E}_x(\omega) = \int_{-\infty}^{\infty} r_x(\tau)e^{-j\omega\tau}\, d\tau. \tag{1.5}$$

Again, we say that $r_x(\tau)$ and $\mathcal{E}_x(\omega)$ constitute a Fourier transform pair:

$$r_x(\tau) \leftrightarrow \mathcal{E}_x(\omega).$$

1.1.2   Discrete-Time Deterministic Signals

Let us suppose that $x(n)$ is a real- or complex-valued deterministic sequence, where n takes integer values, and which is obtained by uniformly sampling the continuous-time signal $x(t)$; read Section 1.2. If $x(n)$ has finite energy $\mathcal{E} = \sum_{n=-\infty}^{\infty} |x(n)|^2 < \infty$, then it has the frequency domain representation (discrete-time Fourier transform)

$$X(\omega) = \sum_{n=-\infty}^{\infty} x(n)e^{-j\omega n},$$

or equivalently

$$X(f) = \sum_{n=-\infty}^{\infty} x(n)e^{-j2\pi f n}.$$

It should be noted that $X(f)$ is periodic with a period of one and $X(\omega)$ is periodic with a period of 2π.

The inverse discrete-time Fourier transform that yields the deterministic sequence $x(n)$ from $X(\omega)$ or $X(f)$ is given by

$$x(n) = \int_{-1/2}^{1/2} X(f)e^{j2\pi f n}\, df = \frac{1}{2\pi} \int_{-\pi}^{\pi} X(\omega)e^{j\omega n}\, d\omega.$$

Notice that the integration limits are related to the periodicity of the spectra.

We refer to $|X(f)|^2$ as the energy density spectrum of $x(n)$ and denote it as

$$\mathcal{E}_x(f) = \left|X(f)\right|^2.$$

The energy density spectrum $\mathcal{E}_x(f)$ of a deterministic discrete-time signal $x(n)$ can also be found by means of the *autocorrelation sequence*

$$r_x(k) = \sum_{n=-\infty}^{\infty} x^*(n)x(n+k)$$

via the discrete-time Fourier transform

$$\mathcal{E}_x(f) = \sum_{k=-\infty}^{\infty} r_x(k)e^{-j2\pi f k}.$$

That is, for a discrete-time signal, the Fourier transform pair is

$$r_x(k) \leftrightarrow \mathcal{E}_x(f).$$

1.1.3   Unit Impulse

In signal analysis a frequently used deterministic signal is the unit impulse. In continuous time the unit impulse $\delta(t)$, also called the delta function, may be defined by the following relation:

$$\int_{-\infty}^{\infty} \delta(t)x(t)\, dt = x(0),$$

where $x(t)$ is an arbitrary signal continuous at $t = 0$. Its area is $\int_{-\infty}^{\infty} \delta(t)\, dt = 1$.

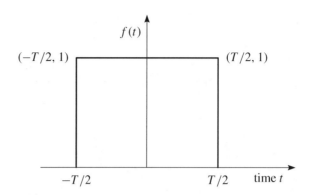

FIGURE 1.1. Rectangular pulse.

In discrete time the unit sample, also called a unit impulse sequence, is defined as

$$\delta(n) = \begin{cases} 1, & n = 0, \\ 0, & n \neq 0. \end{cases}$$

It follows that a continuous-time signal $x(t)$ may be represented as

$$x(t) = \int_{-\infty}^{\infty} x(\tau)\delta(t - \tau)\,d\tau \qquad \text{for all } t.$$

Similarly, a sequence $x(n)$ may be represented as

$$x(n) = \sum_{k=-\infty}^{\infty} x(k)\delta(n - k) \qquad \text{for all } n. \qquad (1.6)$$

The Fourier transform of the unit impulse $\delta(t)$ is given by

$$\int_{-\infty}^{\infty} \delta(t)e^{-j2\pi ft}\,dt = 1,$$

which gives us the following Fourier transform pair:

$$\delta(t) \leftrightarrow 1.$$

The spectrum of the unit sample is obtained by

$$\sum_{n=-\infty}^{\infty} \delta(n)e^{-j2\pi fn} = 1,$$

which gives us the following Fourier transform pair:

$$\delta(n) \leftrightarrow 1.$$

1.1.4 Rectangular Pulse

Let us now consider a single rectangular pulse $f(t)$ with amplitude 1 and pulse width equal to T. In Figure 1.1 we have shifted the pulse $-T/2$ to place it sym-

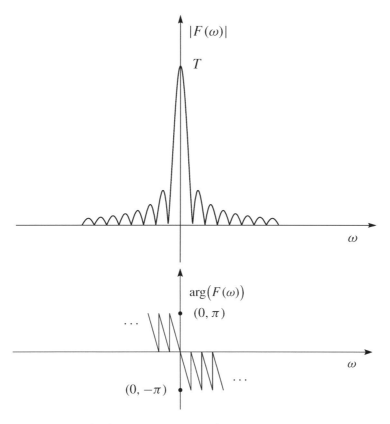

FIGURE 1.2. Top: Magnitude spectrum $|F(\omega)|$ of rectangular pulse. Notice that $F(\omega)$ has zeros at $\pm\frac{2\pi}{T}$, $\pm\frac{4\pi}{T}$, ... Bottom: Phase spectrum $\arg(F(\omega))$ of rectangular pulse. Notice that $\arg(F(\omega))$ has jumps equal to π at $\pm\frac{2\pi}{T}$, $\pm\frac{4\pi}{T}$, ...

metrically around $t = 0$. The equation for the pulse is

$$f(t) = \begin{cases} 1, & |t| \leq T/2, \\ 0, & \text{otherwise.} \end{cases} \tag{1.7}$$

Let the frequency be f in Hz [cycle/s] and $\omega = 2\pi f$ [radian/s]. Then the Fourier transform of $f(t)$ is

$$F(\omega) = T\,\frac{\sin\frac{\omega T}{2}}{\frac{\omega T}{2}} = T\,\text{sinc}\!\left(\frac{\omega T}{2}\right). \tag{1.8}$$

The magnitude spectrum $|F(\omega)|$ and the phase spectrum $\arg(F(\omega))$ are depicted in Figure 1.2. Notice that $\arg(F(\omega))$ of $f(t)$ is linear for $\omega \neq \frac{2\pi n}{T}$ with jumps equal to π for $\omega = \frac{2\pi n}{T}$, because of the change of sign of $\sin\!\left(\frac{\omega T}{2}\right)$ at these points.

From (1.4) follows that the ACF $r_f(\tau)$ for a rectangular pulse has a triangular waveform; see Figure 1.3,

$$r_f(\tau) = \begin{cases} T\!\left(1 - \frac{|\tau|}{T}\right), & \text{for } |\tau| \leq T, \\ 0, & \text{otherwise.} \end{cases} \tag{1.9}$$

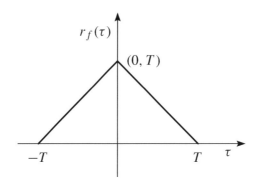

FIGURE 1.3. Autocorrelation function $r_f(\tau)$ of the rectangular pulse shown in Figure 1.1.

The energy density spectrum $\mathcal{E}_f(\omega)$ of $f(t)$ is a real function because $r_f^*(\tau) = r_f(-f)$:

$$\mathcal{E}_f(\omega) = \int_{-\infty}^{\infty} r_f(\tau)e^{-j\omega\tau}\,d\tau = \int_{-\infty}^{\infty} r_f(\tau)\cos(\omega\tau)\,d\tau. \qquad (1.10)$$

The energy density spectrum of the rectangular pulse $f(t)$ is

$$\mathcal{E}_f(\omega) = \int_{-T}^{T} (T - \tau)\cos(\omega\tau)\,d\tau = T^2\left(\frac{\sin\frac{\omega T}{2}}{\frac{\omega T}{2}}\right)^2 = T^2\,\mathrm{sinc}^2\,\frac{\omega T}{2}. \qquad (1.11)$$

The energy density spectrum $\mathcal{E}_f(\omega)$ is depicted in Figure 1.4.

In discrete time the rectangular pulse takes on the form

$$f(n) = \begin{cases} 1, & 0 \le n \le N - 1, \\ 0, & \text{otherwise,} \end{cases} \qquad (1.12)$$

where N is an integer. The Fourier transform of $f(n)$ is

$$F(f) = \sum_{n=0}^{N-1} e^{-j2\pi fn} = \frac{\sin(\pi f N)}{\sin(\pi f)} e^{-j\pi f(N-1)}.$$

1.1.5 Random Signals

A random process can be viewed as a mapping of the outcomes of a random experiment to a set of functions of time—in this context a signal $X(t)$. Such a signal is *stationary* if the density functions $p(X(t))$ describing it are invariant under *translation* of time t. A random stationary process is an infinite energy signal, and therefore its Fourier transform does not exist. The spectral characteristics of a random process is obtained according to the Wiener–Khinchine theorem [see, e.g., Shanmugan & Breipohl (1988)] by computing the Fourier transform of the ACF. That is, the distribution of signal power as a function of frequency is given by

$$S_X(\omega) = \int_{-\infty}^{\infty} r_X(\tau)e^{-j\omega\tau}\,d\tau. \qquad (1.13)$$

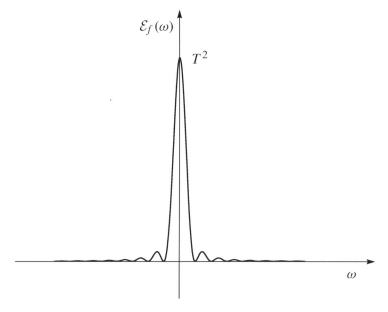

FIGURE 1.4. Energy density spectrum $\mathcal{E}_f(\omega)$ of the rectangular pulse shown in Figure 1.1. Note that $\mathcal{E}_f(\omega)$ has zeros at $\pm\frac{2\pi}{T}, \pm\frac{4\pi}{T}, \pm\frac{6\pi}{T}, \ldots$.

The ACF of the stationary process $X(t)$ is defined as $r_X(\tau) = E\{X(t)^*X(t+\tau)\}$ with $E\{\cdot\}$ denoting the expectation operator and τ being the lag. The *inverse Fourier transform* is given by

$$r_X(\tau) = \int_{-\infty}^{\infty} S_X(\omega)e^{j\omega\tau}\,d\omega. \tag{1.14}$$

The quantity $S_X(\omega)$ is called the *power density spectrum* of $X(t)$.

A discrete-time random process (sequence) has infinite energy but has a finite average power given by $E(X^2(n)) = r_X(0)$. According to the Wiener–Khinchine theorem we obtain the spectral characteristic of the discrete-time random process by means of the Fourier transform of the autocorrelation sequence $r_X(m)$:

$$S_X(f) = \sum_{m=-\infty}^{\infty} r_X(m)e^{-j2\pi fm}.$$

The inverse Fourier transform is

$$r_X(m) = \int_{-1/2}^{1/2} S_X(f)e^{j2\pi fm}\,df.$$

1.1.6   Random Sequence of Pulses

In Section 1.1.4 we studied the characteristics of a single rectangular pulse $f(t)$. Next we want to become familiar with the same properties for a random sequence of pulses with amplitude ±1; each pulse with duration T.

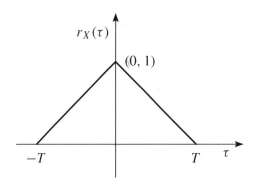

FIGURE 1.5. Autocorrelation function $r_X(\tau)$ for random sequence of pulses with amplitude ± 1.

The ACF for the sample function $x(t)$ of a process $X(t)$ consisting of a random sequence of pulses with amplitude ± 1 and with equal probability for the outcome $+1$ and -1 is [see, e.g., Forssell (1991) or Haykin (2000)]

$$r_X(\tau) = \begin{cases} \left(1 - \frac{|\tau|}{T}\right), & \text{for } |\tau| \le T, \\ 0, & \text{otherwise.} \end{cases} \tag{1.15}$$

The ACF is plotted in Figure 1.5. It follows that the power spectral density is

$$S_X(\omega) = \int_{-T}^{T} \left(1 - \frac{|\tau|}{T}\right) e^{-j\omega\tau} \, d\tau = T \, \text{sinc}^2\left(\frac{\omega T}{2}\right), \tag{1.16}$$

which is plotted in Figure 1.6. The power spectral density of $X(t)$ possesses a main lope bounded by well-defined spectral nulls. Accordingly, the null-to-null bandwidth provides a simple measure for the bandwidth of $X(t)$.

Note that the power spectral density $S_X(\omega)$ of a random sequence of pulses with amplitude ± 1 differs from the energy spectral density $\mathcal{E}_f(\omega)$, given in (1.11), of a single rectangular pulse by only a scalar factor T.

1.2   Sampling

A crucial signal processing operation in a GPS or Galileo software-defined receiver is *sampling*. In the following we briefly review the sampling process.

Consider the signal $x(t)$. Suppose that we sample this signal at a uniform rate— say once every T_s seconds. Then we obtain an infinite sequence of samples, and we denote this sequence by $\{x(nT_s)\}$, where n takes on all integer values. The quantity T_s is called the *sampling period*, and its reciprocal $f_s = 1/T_s$ the *sampling rate*.

The sampling operation is mathematically described by

$$x_\delta(t) = \sum_{n=-\infty}^{\infty} x(nT_s)\delta(t - nT_s),$$

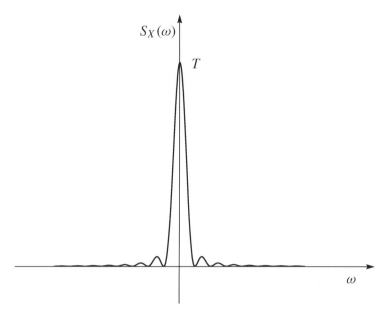

$$S_X(\omega)$$

T

ω

FIGURE 1.6. Power spectral density $S_X(\omega)$ of a random sequence of pulses.

where $x(t)$ is the signal being sampled, and $x_\delta(t)$ is the sampled signal that consists of a sequence of impulses separated in time by T_s. The term $\delta(t - nT_s)$ represents a delta function positioned at time $t = nT_s$. The Fourier transform of $x_\delta(t)$ is

$$X_\delta(f) = f_s \sum_{n=-\infty}^{\infty} X(f - nf_s).$$

Figures 1.7 and 1.8 show the sampling process in the time and frequency domain, respectively.

Figure 1.8 reveals that if the sampling rate f_s is lower that $2B$, then the frequency-shifted components of $X(f)$ overlap and the spectrum of the sampled signal is not similar to the spectrum of the original signal $x(t)$. The spectral overlap effect is known as *aliasing*, and the sampling rate $f_s = 2B$ is called the *Nyquist rate*.

To avoid the effects of aliasing, we may use a lowpass anti-aliasing filter to attenuate frequency components above B, (see Figure 1.8), and sample the signal with a rate higher than the Nyquist rate, i.e., $f_s > 2B$. We return to the issue of sampling in Chapter 4.

1.3   Characterization of Systems

In the continuous-time domain, a system is a functional relationship between the input signal $x(t)$ and the output signal $y(t)$. The input–output relation of a system may be denoted as

$$y(t_0) = f\big(x(t)\big), \quad \text{where } -\infty < t, t_0 < \infty. \tag{1.17}$$

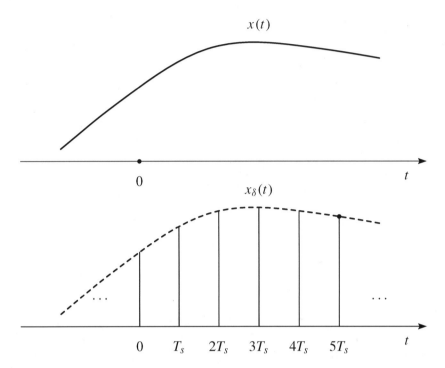

FIGURE 1.7. Sampling operation shown in the time domain. Top: Signal $x(t)$. Bottom: Sampled signal $x_\delta(t)$.

Figure 1.9 shows a block diagram of a system characterized by a function f and with input signal $x(t)$ and output signal $y(t)$.

By means of the properties of the input–output relationship given in (1.17), we can classify systems as follows:

Linear and nonlinear systems A system is said to be linear if superposition applies. That is, if

$$y_1(t) = f\big(x_1(t)\big) \qquad \text{and} \qquad y_2(t) = f\big(x_2(t)\big),$$

then

$$a_1 y_1(t) + a_2 y_2(t) = f\big(a_1 x_1(t) + a_2 x_2(t)\big).$$

A system in which superposition does not apply is termed a nonlinear system.

Time-invariant and time-varying systems A system is said to be be time-invariant if a time shift in the input results in a corresponding time shift in the output. That is,

$$\text{if } y(t) = f\big(x(t)\big), \text{ then } y(t - t_0) = f\big(x(t - t_0)\big) \text{ for } -\infty < t, t_0 < \infty,$$

where t_0 is any real number. Systems that do not meet this requirement are called time-varying systems.

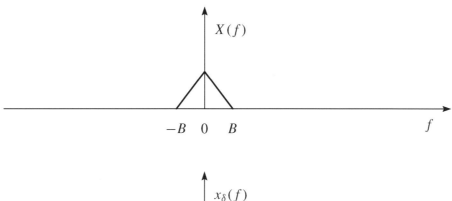

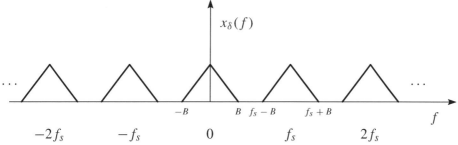

FIGURE 1.8. Sampling operation shown in the frequency domain. Top: Signal $X(f)$ with bandwidth B. Bottom: $x_\delta(f)$ when $f_s > 2B$.

Causal and noncausal systems A system is said to be causal if its response does not begin before the input is applied, or in other words, the value of the output at $t = t_0$ depends only on the values of $x(t)$ for $t \leq t_0$. In mathematical terms, we have

$$y(t_0) = f\big(x(t)\big) \quad \text{for} \quad t \leq t_0 \quad \text{and} \quad -\infty < t, t_0 < \infty.$$

Noncausal systems do not satisfy the condition given above. Moreover, they do not exist in a real world but can be approximated by the use of time delay.

The classification of continuous-time systems easily carries over to discrete-time systems. Here the input and output signals are sequences, and the system maps the input sequence $x(n)$ into the output sequence $y(n)$.

A simple example of a discrete-time linear system is a system that is a linear combination of the present and two past inputs. Such a system can in general be described by

$$y(n) = x(n) + a_1 x(n-1) + a_2 x(n-2) \tag{1.18}$$

and is illustrated in Figure 1.10.

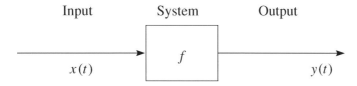

FIGURE 1.9. Block diagram representation of a system.

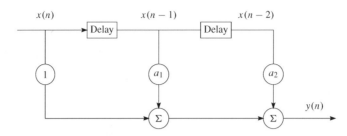

FIGURE 1.10. Simple linear system with input–output relation $y(n) = x(n)+a_1 x(n-1)+ a_2 x(n-2)$.

1.4   Linear Time-Invariant Systems

Let us first consider a continuous-time, linear, time-invariant (LTI) system characterized by an impulse response $h(t)$, which is defined to be the response $y(t)$ from the LTI system to a unit impulse $\delta(t)$. That is,

$$h(t) \equiv y(t) \qquad \text{when} \qquad x(t) = \delta(t).$$

The response to the input $x(t)$ is found by convolving $x(t)$ with $h(t)$ in the time domain:

$$y(t) = x(t) * h(t) = \int_{-\infty}^{\infty} h(\lambda)x(t-\lambda)d\lambda. \qquad (1.19)$$

Convolution is denoted by the $*$. Since $h(t) = 0$ for $t < 0$ for causal systems, we can also write $y(t)$ as

$$y(t) = \int_{-\infty}^{\infty} x(\lambda)h(t-\lambda)\, d\lambda.$$

Define the continuous-time exponential signal $x(t) = e^{j2\pi ft}$. From (1.19) we then have

$$y(t) = h(t) * x(t) = h(t) * e^{j2\pi ft} = \int_{-\infty}^{\infty} h(\lambda)e^{j2\pi f(t-\lambda)}\, d\lambda$$

$$= e^{j2\pi ft}\left[\int_{-\infty}^{\infty} h(\lambda)e^{-j2\pi f\lambda}\, d\lambda\right]. \qquad (1.20)$$

The expression in brackets is the Fourier transform of $h(t)$, which we denote $H(f)$.

Now define the discrete-time exponential sequence $x(n) = e^{j2\pi fn} = e^{j\omega n}$. We can then characterize a discrete-time LTI system by its frequency response to $x(n)$. By means of the convolution sum formula, a discrete version of (1.20), we obtain the response

$$y(n) = \sum_{k=-\infty}^{\infty} e^{j\omega k} h(n-k) = \left(\sum_{k=-\infty}^{\infty} h(k)e^{-j\omega k}\right) e^{j\omega n}. \qquad (1.21)$$

The expression in parentheses is the discrete-time Fourier transform of the impulse response $h(n)$, which we denote $H(\omega)$.

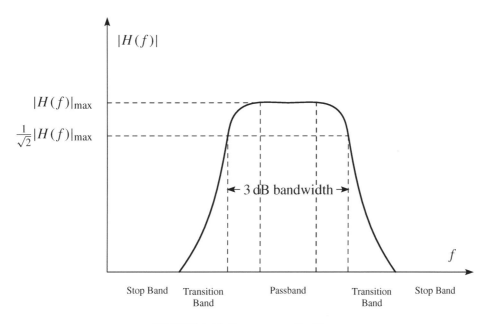

FIGURE 1.11. Parameters of a filter.

The convolution in the time domain corresponds to the multiplication of Fourier transforms in the frequency domain. Thus, for the system under consideration the Fourier transforms of the input and output signals are related to each other by

$$Y(f) = H(f)X(f). \tag{1.22}$$

In general, the transfer function $H(f)$ is a complex quantity and can be expressed in magnitude and angle form as

$$H(f) = |H(f)|e^{j \arg(H(f))}.$$

The quantity $|H(f)|$ is called the amplitude response of the system, and the quantity $\arg(H(f))$ is called the phase response of the system. The magnitude response is often expressed in decibels (dB) using the definition

$$|H(f)|_{dB} = 20 \log_{10}|H(f)|.$$

We mention in passing that in real systems $h(t)$ is a real-valued function and hence $H(f)$ has conjugate symmetry in the frequency domain, i.e., $H(f) = H^*(-f)$.

If the input and output signals are expressed in terms of power spectral density, then the input–output relation is given by

$$S_y(f) = |H(f)|^2 S_x(f).$$

The equations above show that an LTI system acts as a filter. Filters can be classified into lowpass, bandpass, and highpass filters and they are often characterized by stopbands, passband, and half-power (3 dB) bandwidth. These parameters are identified in Figure 1.11 for a bandpass filter.

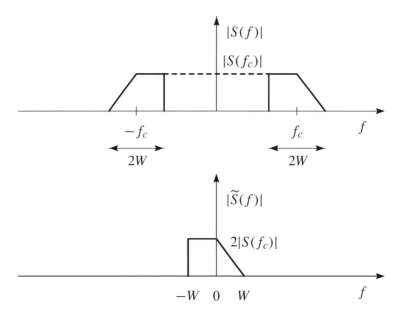

FIGURE 1.12. Magnitude spectrum of bandpass signal $s(t)$ and complex envelope $\tilde{s}(t)$.

1.5   Representation of Bandpass Signals

Signals can also be classified into lowpass, bandpass, and highpass categories depending on their spectra. We define a bandpass signal as a signal with frequency content concentrated in a band of frequencies above zero frequency. Bandpass signals arise in the GPS and Galileo systems where the information-bearing signals are transmitted over bandpass channels from the satellite to the receiver.

Let us now consider an analog signal $s(t)$ with frequency content limited to a narrow band of total extent $2W$ and centered about some frequency $\pm f_c$; see Figure 1.12. We term such a signal a *bandpass signal* and represent it as

$$s(t) = a(t) \cos\big(2\pi f_c t + \varphi(t)\big), \tag{1.23}$$

where $a(t)$ is called the *amplitude* or *envelope of the signal* and $\varphi(t)$ is called the phase of the signal. The frequency f_c is called the carrier frequency. Equation (1.23) represents a hybrid form of amplitude modulation and angle modulation and it includes amplitude modulation, frequency modulation, and phase modulation as special cases. For more details on this topic we refer to Haykin (2000).

By expanding the cosine function in (1.23) we obtain an alternative representation of the bandpass signal:

$$\begin{aligned} s(t) &= a(t) \cos\big(\varphi(t)\big) \cos(2\pi f_c t) - a(t) \sin\big(\varphi(t)\big) \sin(2\pi f_c t) \\ &= s_I(t) \cos(2\pi f_c t) - s_Q(t) \sin(2\pi f_c t), \end{aligned} \tag{1.24}$$

where by definition

$$s_I(t) = a(t)\cos(\varphi(t)),\tag{1.25}$$

$$s_Q(t) = a(t)\sin(\varphi(t)).\tag{1.26}$$

Since the sinusoids $\cos(2\pi f_c t)$ and $\sin(2\pi f_c t)$ differ by $90°$, we say that they are in *phase quadrature*. We refer to $s_I(t)$ as the in-phase component of $s(t)$ and to $s_Q(t)$ as the quadrature component of the bandpass signal $s(t)$. Both $s_I(t)$ and $s_Q(t)$ are real-valued lowpass signals.

Another representation of the bandpass signal $s(t)$ is obtained by defining the complex envelope

$$\widetilde{s}(t) = s_I(t) + js_Q(t)\tag{1.27}$$

such that

$$s(t) = \Re\big(\widetilde{s}(t)e^{j2\pi f_c t}\big).\tag{1.28}$$

Equation (1.27) can be seen as the Cartesian form of expressing the complex envelope $\widetilde{s}(t)$. In polar form we may write $\widetilde{s}(t)$ as

$$\widetilde{s}(t) = a(t)e^{j\varphi(t)},\tag{1.29}$$

where $a(t) = \sqrt{s_I^2(t) + s_Q^2(t)}$ and $\varphi(t) = \arctan\left(\frac{s_Q(t)}{s_I(t)}\right)$ are both real-valued lowpass signals. That is, the information carried in the bandpass signal $s(t)$ is preserved in $\widetilde{s}(t)$ whether we represent $s(t)$ in terms of its in-phase and quadrature components as in (1.24) or in terms of its envelope and phase as in (1.23).

In the frequency domain the bandpass signal $s(t)$ is represented by its Fourier transform

$$\begin{aligned}
S(f) &= \int_{-\infty}^{\infty} s(t)e^{-j2\pi ft}\, dt\\
&= \int_{-\infty}^{\infty} \Re\big(\widetilde{s}(t)e^{j2\pi f_c t}\big)e^{-j2\pi ft}\, dt\\
&= \frac{1}{2}\int_{-\infty}^{\infty}\big(\widetilde{s}(t)e^{j2\pi f_c t} + \widetilde{s}^*(t)e^{-j2\pi f_c t}\big)e^{-j2\pi ft}\, dt\\
&= \frac{1}{2}\int_{-\infty}^{\infty}\widetilde{s}(t)e^{-j2\pi(f-f_c)t}\, dt + \frac{1}{2}\int_{-\infty}^{\infty}\widetilde{s}^*(t)e^{-j2\pi(f+f_c)t}\, dt.
\end{aligned}\tag{1.30}$$

In the above derivation we used $\Re(v) = \frac{1}{2}(v + v^*)$. If we denote the Fourier transform of the signals $s(t)$ and $\widetilde{s}(t)$ as $S(f)$ and $\widetilde{S}(f)$, respectively, we may write (1.30) as

$$S(f) = \frac{1}{2}\big(\widetilde{S}(f - f_c) + \widetilde{S}^*(-f - f_c)\big).\tag{1.31}$$

This relation is illustrated in Figure 1.12.

2
GPS Signal

In order to design a software-defined single frequency GPS receiver it is necessary to know the characteristics of the signal and data transmitted from the GPS satellites and received by the GPS receiver antenna. In this chapter an overview of the GPS signal generation scheme and the most important properties of the various signals and data are presented.

2.1 Signals and Data

The GPS signals are transmitted on two radio frequencies in the UHF band. The UHF band covers the frequency band from 500 MHz to 3 GHz. These frequencies are referred to as L1 and L2 and are derived from a common frequency, $f_0 = 10.23$ MHz:

$$f_{L1} = 154 f_0 = 1575.42 \text{ MHz}, \qquad (2.1)$$
$$f_{L2} = 120 f_0 = 1227.60 \text{ MHz}. \qquad (2.2)$$

The signals are composed of the following three parts:

Carrier The carrier wave with frequency f_{L1} or f_{L2},

Navigation data The navigation data contain information regarding satellite orbits. This information is uploaded to all satellites from the ground stations in the GPS Control Segment. The navigation data have a bit rate of 50 bps. More details on the navigation data can be seen in Section 2.6.

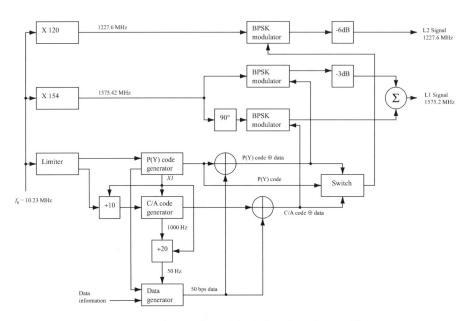

FIGURE 2.1. Generation of GPS signals at the satellites.

Spreading sequence Each satellite has two unique spreading sequences or codes. The first one is the coarse acquisition code (C/A), and the other one is the encrypted precision code (P(Y)). The C/A code is a sequence of 1023 chips. (A chip corresponds to a bit. It is simply called a chip to emphasize that it does not hold any information.) The code is repeated each ms giving a chipping rate of 1.023 MHz. The P code is a longer code ($\approx 2.35 \cdot 10^4$ chips) with a chipping rate of 10.23 MHz. It repeats itself each week starting at the beginning of the GPS week which is at Saturday/Sunday midnight. The C/A code is only modulated onto the L1 carrier while the P(Y) code is modulated onto both the L1 and the L2 carrier. Section 2.3 describes the generation and properties of the spreading sequences in detail.

2.2 GPS Signal Scheme

In the following a detailed description of the signal generation is given. Figure 2.1 is a block diagram describing the signal generation; see Kaplan & Hegarty (2006), page 124.

The block diagram should be read from left to right. At the far left, the main clock signal is supplied to the remaining blocks. The clock signal has a frequency of 10.23 MHz. Actually, the exact frequency is 10.22999999543 MHz to adjust for relativistic effects giving a frequency of 10.23 MHz seen from the user on Earth. When multiplied by 154 and 120, it generates the L1 and L2 carrier signals, respectively. At the bottom left corner a limiter is used to stabilize the clock signal before supplying it to the P(Y) and C/A code generators. At the very bottom the data generator generates the navigation data. The code generators and the

TABLE 2.1.			TABLE 2.2.		
Output of the exclusive OR operation			Output of ordinary multiplication		
Input	**Input**	**Output**	**Input**	**Input**	**Output**
0	0	0	-1	-1	1
0	1	1	-1	1	-1
1	0	1	1	-1	-1
1	1	0	1	1	1

data generator are synchronized through the X_1 signal supplied by the P(Y) code generator.

After code generation, the codes are combined with the navigation data through modulo-2 adders. The exclusive OR operation is used on binary sequences represented by 0's and 1's, and its properties are shown in Table 2.1.

If the binary sequences were represented by the polar non-return-to-zero representation, i.e., 1's and -1's, ordinary multiplication could be used instead. The corresponding properties of the multiplication with two binary non-return-to-zero sequences are shown in Table 2.2.

The *C/A code $\oplus$ data* and the *P(Y) code $\oplus$ data* signals are supplied to the two modulators for the L1 frequency. Here the signals are modulated onto the carrier signal using the binary phase shift keying (BPSK) method. Note that the two codes are modulated in-phase and quadrature with each other on L1. That is, there is a 90° phase shift between the two codes. We return to this issue shortly. After the P(Y) part is attenuated 3 dB, these two L1 signals are added to form the resulting L1 signal. The so-called standard positioning service (SPS) is based on C/A code signals alone.

It follows that the signal transmitted from satellite k can be described as

$$
\begin{aligned}
s^k(t) = \sqrt{2P_C}\big(C^k(t) \oplus D^k(t)\big)\cos(2\pi f_{L1}t) \\
+ \sqrt{2P_{PL1}}\big(P^k(t) \oplus D^k(t)\big)\sin(2\pi f_{L1}t) \\
+ \sqrt{2P_{PL2}}\big(P^k(t) \oplus D^k(t)\big)\sin(2\pi f_{L2}t), \quad (2.3)
\end{aligned}
$$

where P_C, P_{PL1}, and P_{PL2} are the powers of signals with C/A or P code, C^k is the C/A code sequence assigned to satellite number k, P^k is the P(Y) code sequence assigned to satellite number k, D^k is the navigation data sequence, and f_{L1} and f_{L2} are the carrier frequencies of L1 and L2, respectively.

Figure 2.2 shows the three parts forming the signal on the L1 frequency. The C/A code repeats itself every ms, and one navigation bit lasts 20 ms. Hence *for each navigation bit, the signal contains 20 complete C/A codes.*

Figure 2.3 shows the Gold code C, the navigation data D, the modulo-2 added signal $C \oplus D$, and the carrier. The final signal is created by *binary phase-shift keying* (BPSK) where the carrier is instantaneously phase shifted by 180° at the time of a chip change. When a navigation data bit transition occurs (about one third from the right edge), the phase of the resulting signal is also phase-shifted 180°.

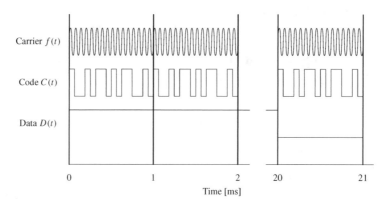

FIGURE 2.2. L1 signal structure: $f(t)$ is the carrier wave and $C(t)$ is the discrete C/A code sequence. As seen, this signal repeats itself every ms. $D(t)$ is the discrete navigation data bit stream. One navigation bit lasts 20 ms. The three parts of the L1 signal are multiplied to form the resulting signal. This figure is not to scale but is only used for illustrative purpose.

The GPS C/A spectrum is illustrated in Figure 3.4.

In summary: For GPS the code length is 1023 chips, 1.023 MHz chipping rate (1 ms period time), 50 Hz data rate (20 code periods per data bit), $\sim 90\%$ of signal power within ~ 2 MHz bandwidth.

2.3   C/A Code

In this section, the spreading sequences used in GPS are described. We restrict ourselves to the C/A code sequences, as we deal only with L1 signals in this book. The spreading sequences used as C/A codes in GPS belong to a unique family of sequences. They are often referred to as *Gold codes*, as Robert Gold described them in 1967; see Gold (1967). They are also referred to as pseudo-random noise sequences, or simply PRN sequences, because of their characteristics.

2.3.1   Gold Sequence

The pseudorandom noise (PRN) codes transmitted by the GPS satellites are deterministic sequences with noiselike properties. Each C/A code is generated using a tapped linear feedback shift register (LFSR); cf. Strang & Borre (1997), Section 14.1. It generates a maximal-length sequence of length $N = 2^n - 1$ elements.

A Gold code is the sum of two maximum-length sequences. The GPS C/A code uses $n = 10$. The sequence $p(t)$ repeats every ms so the chip length is 1 ms$/1023 = 977.5$ ns $\approx 1\,\mu$s, which corresponds to a metric length of 300 m when propagating through vacuum or air. For further details on the generation of the Gold codes, we refer to Section 2.3.3. The ACF for this C/A code is

$$r_p(\tau) = \frac{1}{NT_c} \int_0^{NT_c} p(t)p(t+\tau)\,dt.$$

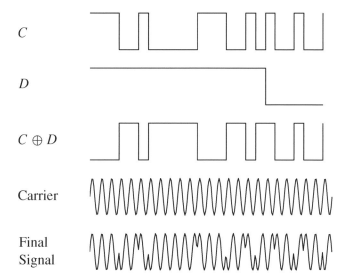

FIGURE 2.3. The effect of BPSK modulation of the L1 carrier wave with the C/A code and the navigation data for one satellite. The resulting L1 signal is the product of G, N, and the carrier signals. The plot contains the first 25 chips of the Gold code for PRN 1.

The sequence would have 512 ones and 511 zeros, and these would appear to be distributed at random. Yet the string of chips so generated is entirely deterministic. The sequence is pseudorandom, not random. Outside the correlation interval the ACF of $p(t)$ is $-1/N$. For the C/A code the constant term is $-1/N = -1/1023$, which is shown in Figure 2.4.

The ACF can be expressed as the sum of this constant term and an infinite series of the triangle function $r_X(\tau)$ defined in (1.15). This infinite series is obtained by the convolution of $r_X(\tau)$ with an infinite series of impulse functions that are phase-shifted by mNT_c:

$$r_P(\tau) = -\frac{1}{N} + \frac{N+1}{N} r_X(\tau) * \sum_{m=-\infty}^{\infty} \delta(\tau + mNT_c), \qquad (2.4)$$

where $*$ denotes convolution. The power (line) spectrum of this periodic PRN sequence is derived from the Fourier transform of (2.4):

$$S_P(\omega) = \frac{1}{N^2}\left(\delta(\omega) + \sum_{\substack{m=-\infty \\ m \neq 0}}^{\infty} (N+1)\operatorname{sinc}^2\left(\frac{m\pi}{N}\right)\delta\left(\omega + \frac{m2\pi}{NT_c}\right)\right) \qquad (2.5)$$

for $m = \pm 1, \pm 2, \pm 3, \ldots$; see Kaplan & Hegarty (2006), page 119.

2.3.2 Gold Sequence Generation—Overview

The generation of the Gold codes is sketched in Figure 2.5. The C/A code generator contains two shift registers known as G_1 and G_2. These shift registers each

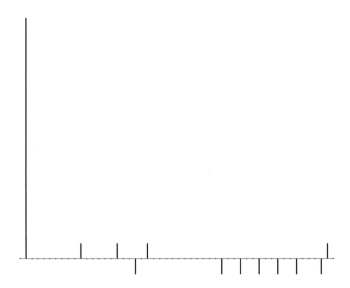

FIGURE 2.4. Stem plot of an ACF for a Gold sequence. The left stem has correlation value $r_p(0) = 1$; all other correlation values are $\frac{63}{1023}$, $\frac{-1}{1023}$, or $\frac{-65}{1023}$. Only the first 50 lags out of 1023 are shown.

have 10 cells generating sequences of length 1023. The two resulting 1023 chip-long sequences are modulo-2 added to generate a 1023 chip-long C/A code, only if the polynomial is able to generate code of maximum length.

Every 1023rd period, the shift registers are reset with all ones, making the code start over. The G_1 register always has a feedback configuration with the polynomial

$$f(x) = 1 + x^3 + x^{10}, \tag{2.6}$$

meaning that state 3 and state 10 are fed back to the input. In the same way, the G_2 register has the polynomial

$$f(x) = 1 + x^2 + x^3 + x^6 + x^8 + x^9 + x^{10}. \tag{2.7}$$

To make different C/A codes for the satellites, the output of the two shift registers are combined in a very special manner. The G_1 register always supplies its output, but the G_2 register supplies two of its states to a modulo-2 adder to generate its output. The selection of states for the modulo-2 adder is called the phase selection. Table 2.3 shows the combination of the phase selections for each C/A code. It also shows the first 10 chips of each code in octal representation.

As the generation of the C/A codes are of immense importance, we outline in detail the principle of operation of the C/A code generator in the next section.

2.3.3   Gold Sequence Generation—Details

A shift register is a set of one bit storage or memory cells. When a clock pulse is applied to the register, the content of each cell shifts one bit to the right. The

TABLE 2.3. C/A code phase assignment. The selection of different states for the code phase generates the different C/A codes for the GPS satellites.

Satellite ID number	GPS PRN signal number	Code phase selection G_2	Code delay chips	First 10 chips octal
1	1	$2 \oplus 6$	5	1440
2	2	$3 \oplus 7$	6	1620
3	3	$4 \oplus 8$	7	1710
4	4	$5 \oplus 9$	8	1744
5	5	$1 \oplus 9$	17	1133
6	6	$2 \oplus 10$	18	1455
7	7	$1 \oplus 8$	139	1131
8	8	$2 \oplus 9$	140	1454
9	9	$3 \oplus 10$	141	1626
10	10	$2 \oplus 3$	251	1504
11	11	$3 \oplus 4$	252	1642
12	12	$5 \oplus 6$	254	1750
13	13	$6 \oplus 7$	255	1764
14	14	$7 \oplus 8$	256	1772
15	15	$8 \oplus 9$	257	1775
16	16	$9 \oplus 10$	258	1776
17	17	$1 \oplus 4$	469	1156
18	18	$2 \oplus 5$	470	1467
19	19	$3 \oplus 6$	471	1633
20	20	$4 \oplus 7$	472	1715
21	21	$5 \oplus 8$	473	1746
22	22	$6 \oplus 9$	474	1763
23	23	$1 \oplus 3$	509	1063
24	24	$4 \oplus 6$	512	1706
25	25	$5 \oplus 7$	513	1743
26	26	$6 \oplus 8$	514	1761
27	27	$7 \oplus 9$	515	1770
28	28	$8 \oplus 10$	516	1774
29	29	$1 \oplus 6$	859	1127
30	30	$2 \oplus 7$	860	1453
31	31	$3 \oplus 8$	861	1625
32	32	$4 \oplus 9$	862	1712
—	33	$5 \oplus 10$	863	1745
—	34	$4 \oplus 10$	950	1713
—	35	$1 \oplus 7$	947	1134
—	36	$2 \oplus 8$	948	1456
—	37	$4 \oplus 10$	950	1713

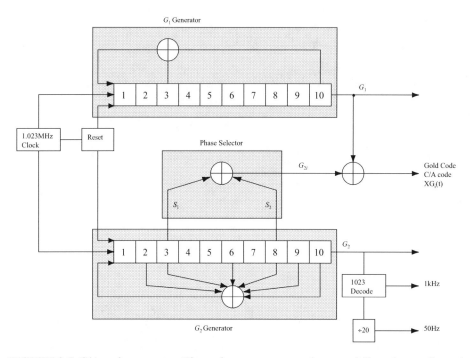

FIGURE 2.5. C/A code generator. The code generator contains two shift registers, G_1 and G_2. The output from G_2 depends on the phase selector. The different configurations of the phase selector makes the different C/A codes.

content of the last cell is "read out" as output. The special properties of such shift registers depend on how information is "read in" to cell 1.

For a tapped linear feedback shift register, the input to cell 1 is determined by the state of the other cells. For example, the binary sum from cells 3 and 10 in a 10-cell register could be the input. If cells 3 and 10 have different states (one is 1 and the other 0), a 1 will be read into cell 1 on the next clock pulse. If cells 3 and 10 have the same state, 0 will be read into cell 1. If we start with 1 in every cell, 12 clock pulses later the contents will be 0010001110. The next clock pulse will take the 1 in cell 3 and the 0 in cell 10 and place their sum (1) in cell 1. Meanwhile, all other bits have shifted cell to the right, and the 0 in cell 10 becomes the next bit in the output. A shorthand way of denoting this particular design is by the modulo-2 polynomial $f(x) = 1 + x^3 + x^{10}$. Such a polynomial representation is particularly useful because if $1/f(x) = h_0 + h_1 x + h_2 x^2 + h_3 x^3 + \cdots$, then the coefficients $h_0, h_1, h_2, \ldots$ form the binary output sequence.

The C/A code is generated by two 10-bit LFSRs of maximal length $2^{10} - 1$. One is the $1 + x^3 + x^{10}$ register already described and is referred to as G_1. The other has $f(x) = 1 + x^2 + x^3 + x^6 + x^8 + x^9 + x^{10}$. Cells 2, 3, 6, 8, 9, and 10 are tapped and binary-added to get the new input to cell 1. In this case, the output comes not from cell 10 but from a second set of taps. Various pairs of these second taps are binary-added. The different pairs yield the same sequence with different delays or shifts (as given by the "shift and add" or "cycle and add" property: a chip-by-chip sum of a maximal-length register sequence and any shift of itself is the same sequence except for a shift). The delayed version of the G_2 sequence is binary-added to the output of G_1. That becomes the C/A code. The G_1 and G_2 shift registers are set to the all-ones state in synchronism with the epoch of the X_1 code used in the generation of the P code (see ahead). The various alternative pairs of G_2 taps (delays) are used to generate the complete set of 36 unique PRN C/A codes. These are Gold codes, Gold (1967), Dixon (1984), and any two have a very low cross correlation (are nearly orthogonal).

There are actually 37 PRN C/A codes, but two of them (34 and 37) are identical. A subset of the first 32 codes are assigned to (nominally 24) satellites and recycled when old satellites die and new satellites are launched. Codes 33 through 37 are reserved for other uses, including ground transmitters.

The P code generation follows the same principles as the C/A code, except that 4 shift registers with 12 cells are used. Two registers are combined to produce the X_1 code, which is 15,345,000 chips long and repeats every 1.5 seconds; and two registers are combined to produce the X_2 code, which is 15,345,037 chips long. The X_1 and X_2 codes can be combined with 37 different delays on the X_2 code to produce 37 different one-week segments of the P code. Each of the first 32 segments is associated with a different satellite.

2.3.4 Correlation Properties

The Gold codes are selected as spreading sequences for the GPS signals because of their characteristics. The most important characteristics of the C/A codes are their correlation properties. These properties are described now.

The two important correlation properties of the C/A codes can be stated as follows:

Nearly no cross correlation All the C/A codes are nearly uncorrelated with each other. That is, for two codes C^i and C^k for satellites i and k, the cross correlation can be written as

$$r_{ik}(m) = \sum_{l=0}^{1022} C^i(l)C^k(l+m) \approx 0 \qquad \text{for all } m. \qquad (2.8)$$

Nearly no correlation except for zero lag All C/A are nearly uncorrelated with themselves, except for zero lag. This property makes it easy to find out when two similar codes are perfectly aligned. The autocorrelation property for satellite k can be written as

$$r_{kk}(m) = \sum_{l=0}^{1022} C^k(l)C^k(l+m) \approx 0 \qquad \text{for } |m| \geq 1. \qquad (2.9)$$

Figure 2.6 shows an example of the auto- and cross-correlation properties of the C/A code. As expected, the figure shows high correlation at lag 0 when correlating with the same C/A code, and low correlation when correlating with another C/A code.

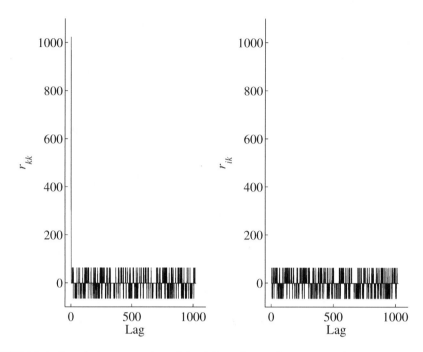

FIGURE 2.6. Correlation properties of the C/A codes. Left: Autocorrelation $r_{kk}(n)$ of the C/A code for PRN 1. Right: Cross correlation $r_{ik}(n)$ of the C/A codes for PRNs 1 and 2.

The autocorrelation shown in the left part of Figure 2.6 has a peak of magnitude

$$r_{kk,\text{peak}} = 2^n - 1 = 1023, \tag{2.10}$$

where n is the number of states in the shift registers. In this case, n equals 10. The remaining values satisfy the following inequality, Gold (1967):

$$|r_{kk}| \leq 2^{(n+2)/2} + 1. \tag{2.11}$$

For $n = 10$, we get

$$|r_{kk}| \leq 65. \tag{2.12}$$

The cross correlation in the right part of Figure 2.6 also satisfies the inequality in (2.11).

2.4   Doppler Frequency Shift

In GPS we are faced with a Doppler frequency shift caused by the motion of the transmitter (satellite) relative to the GPS receiver. The Doppler frequency shift affects both the acquisition and tracking of the GPS signal. For a stationary GPS receiver the maximum Doppler frequency shift for the L1 frequency is around $\pm 5\,\text{kHz}$ and for a GPS receiver moving at high speed it is reasonable to assume that the maximum Doppler shift is $\pm 10\,\text{kHz}$.

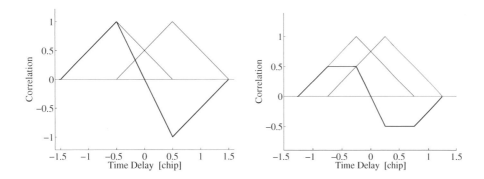

FIGURE 2.7. The two triangles indicate the early and the late ACF. In the left part the two triangles are separated by $d = 1$ chip (classical wide correlator), and the right part shows a separation of $d = 0.5$ chip (narrow correlator). Both discriminators have the same slope close to the origin.

The Doppler frequency shift on the C/A code is small because of the low chip rate of the C/A code. The C/A code has a chip rate of 1.023 Mhz, which is $1575.42/1.023 = 1540$ times lower than the L1 carrier frequency. It follows that the *Doppler frequency on the C/A code is 3.2 Hz and 6.4 Hz for the stationary and the high-speed GPS receiver*, respectively.

The Doppler frequency on the C/A code can cause misalignment between the received and the locally generated codes and the values of the Doppler frequency are important for the tracking method. We return to this topic in Chapters 6 and 7.

2.5   Code Tracking

GPS signal receiving involves a classical problem, namely that one of *code tracking* which can be solved by means of the delay-locked loop (DLL) scheme based on an early–late discriminator. For more detail on this topic we refer to Chapter 7. This loop-error detector or *loop discriminator* is—in its simplest version—based on a two-correlator structure. *Each correlator is set with a small time offset relative to the promptly received signal code timing phase*, both producing early and late signals. Combining the early and late signals provides an error signal, driving the loop toward elimination of the delay-tracking error.

The code tracking loop can be designed as a

– coherent delay lock loop (DLL), or a

– noncoherent DLL.

The coherent DLL requires the phase lock loop to be in lock, i.e., it is tracking the carrier phase and the navigation data bits have been removed from the signal. The noncoherent DLL is able to track the code with the navigation data bit present and the PLL is not necessarily in lock. The noncoherent design is the preferred one; see Winkel (2000).

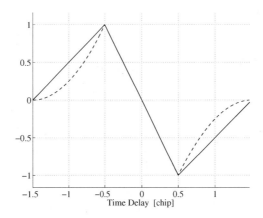

FIGURE 2.8. Coherent (full line) and noncoherent (dashed line) discriminator functions.

The input signal is split into two paths and correlated with two versions, an early and a late, of a locally generated PRN code. The two versions are equally spaced, typically ± 0.5 chip, about the prompt PRN code. This is done for both the in-phase I branch and the quadrature Q branch. The estimation of the ACF in all six signals $I_E, I_L, I_P, Q_E, Q_L, Q_P$, where E denotes early, L denotes late, and P denotes promptly, is based on a summation of the respective signals over an interval of time T. Often T equals $\frac{1}{50}$ s corresponding to the bit duration of the navigation bits. In this time interval the data bit is constant. The GPS signal structure implies that the duration of one navigation data bit divided by one chip duration is an integer (20,460).

The ACF for the signals discussed above can be used to set up the corresponding code discriminators. The early minus late discriminator is obtained by subtracting a late copy of the correlation function from an early copy. The correlator spacing d between the early and the late code is set to 0.5 chip; see Figure 2.7.

The code tracking loop is a device that estimates the signed time difference between the received and the reference code. The zero of this detector is the *pseudorange* or code phase observation. Null tracking is enabled by subtracting the late correlation from the early correlation. The resulting difference is called the *discriminator function*. For a coherent DLL we have

$$D(t) = R\left(t + \tfrac{d}{2}\right) - R\left(t - \tfrac{d}{2}\right).$$

For a noncoherent DLL, the discriminator function is

$$D(t) = R^2\left(t + \tfrac{d}{2}\right) - R^2\left(t - \tfrac{d}{2}\right);$$

see Figure 2.8.

2.6   Navigation Data

The navigation data are transmitted on the L1 frequency with the earlier mentioned bit rate of 50 bps. This section describes the structure and contents of the

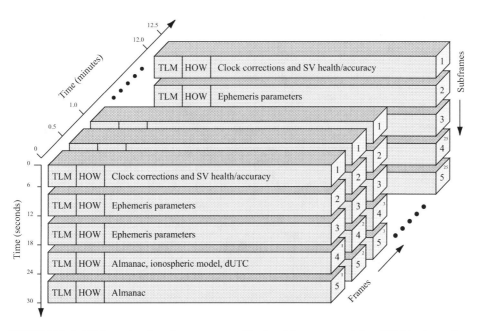

FIGURE 2.9. GPS navigation data structure. Each subframe containing 300 bits lasts 6 s. Subframes 1, 2, and 3 repeat every 30 s while subframes 4 and 5 have 25 versions before repeating. That is, the entire navigation message repeats after 12.5 minutes. Courtesy: Frank van Diggelen.

navigation data. Figure 2.9 shows the overall structure of an entire navigation message.

The basic format of the navigation data is a 1500-bit-long frame containing 5 subframes, each having length 300 bits. One subframe contains 10 words, each word having length 30 bits. Subframes 1, 2, and 3 are repeated in each frame. The last subframes, 4 and 5, have 25 versions (with the same structure, but different data) referred to as page 1 to 25. With the bit rate of 50 bps, the transmission of a subframe lasts 6 s, one frame lasts 30 s, and one entire navigation message lasts 12.5 minutes.

2.6.1 Telemetry and Handover Words

The subframes of 10 words always begin with two special words, the telemetry (TLM) and handover word (HOW) pair.

TLM is the first word of each subframe and it is thus repeated every 6 s. It contains an 8-bit preamble followed by 16 reserved bits and parity. The preamble should be used for frame synchronization.

HOW contains a 17-bit truncated version of the *time of week* (TOW), followed by two flags supplying information to the user of antispoofing, etc. The next three bits indicate the subframe ID to show in which of the five subframes in the current frame this HOW is located.

2.6.2   Data in Navigation Message

In addition to the TLM and HOW words, each subframe contains eight words of data. This will only be a cursory description of the data in the different words and not a complete description of all bits.

Subframe 1 – Satellite Clock and Health Data The first subframe contains first of all clock information. That is information needed to compute at what time the navigation message is transmitted from the satellite. Additionally, subframe 1 contains health data indicating whether or not the data should be trusted.

Subframes 2 and 3 – Satellite Ephemeris Data Subframes 2 and 3 contain the satellite ephemeris data. The ephemeris data relate to the satellite orbit and are needed to compute a satellite position.

Subframes 4 and 5 – Support Data As mentioned, the last two subframes repeat every 12.5 minutes, giving a total of 50 subframes. Subframes 4 and 5 contain almanac data. The almanac data are the ephemerides and clock data with reduced precision. Additionally, each satellite transmits almanac data for all GPS satellites while it only transmits ephemeris data for itself. The remainder of subframes 4 and 5 contain various data, e.g., UTC parameters, health indicators, and ionospheric parameters.

For a more in-depth description of the contents of the navigation data, see SPS (1995).

3
Galileo Signal

The Galileo system offers several services, a few are free of charge and the rest are commercial. In this book we deal only with the L1 OS signal (OS for open service).

The L1 OS signal alone is expected to guarantee a horizontal accuracy better than 15 m, a vertical accuracy better than 35 m, a velocity accuracy better than 50 cm/s, and a timing accuracy better than 100 ns.

The augmentation from a single-frequency to a dual-frequency Galileo receiver naturally includes the E5a signal. For a dual frequency receiver the corresponding accuracies are 7 m, 15 m, 20 cm/s, and 100 ns.

All Galileo satellites use the same frequency bands and make use of code division multiple access (CDMA) technique. Spread spectrum signals will be transmitted including different ranging codes per signal, per frequency, and per satellite. All signals are transmitted in a right-hand circular polarization.

In this chapter we rely on Anonymous (2005). For more details the reader is referred to this document; also note that all information is to be considered preliminary.

3.1 Signal Theoretical Considerations

Today when designing the Galileo signals the situation is very different from the days when the GPS signals were designed. Nowadays applications with difficult signal reception set the specifications for GNSS; the receiver may be used in the woods or indoors. This puts the most demanding efforts on the signal design.

Additionally, the wealth of digital signal processing experience over that time means many hands to the pump. The following considerations are based on Mattos (2004).

The L1 OS signal is transmitted on the frequency $f_1 = 1575.42\,\text{MHz}$. The signal is composed of *three channels*, called A, B, and C. L1-A is identical to L1 PRS (PRS for public regulated service), which is a restricted access signal. Its ranging codes and navigation data are encrypted. The data signal is L1-B (meaning the B channel within L1) and the data-free signal is L1-C (meaning the C channel within L1). A data-free signal is also called a pilot signal. It is made of a ranging code only, not modulated by a navigation data stream.

The L1 OS signal has a 4092 code length with a 1.023 MHz chipping rate giving it a repetition rate of 4 ms; on the pilot signal a secondary code of length 25 chips extends the repetition interval to 100 ms.

Under some circumstances it may be difficult to separate the wanted signal from the unwanted ones and the unwanted one is often a cross correlation from another satellite as the inherent CDMA isolation of the codes is only around 21 dB. The cross-correlation problem is solved by using *very long codes*. However, longer codes also delay the acquisition process. In most cases the processor must search at half-chip offsets; thus, 8184 possibilities for the L1 OS code. To search the very long code lengths proposed for the new signals would be impractical, so the codes have been designed with escape routes. The most common one is called a *tiered code*. This means it is built in layers so that when you have a strong signal you can acquire on a simple layer, with less time-domain possibilities, only switching to the full-length code when required.

The minimum *bandwidth* is generally twice the chipping rate for simple codes, while for BOC codes it is twice the sum of chipping rate and offset code rate. Thus, the minimum practical bandwidth for the Galileo L1 OS is 8 MHz. For precise tracking of the code a bandwidth wider than the minimum is generally used.

Within this 4 ms period the signal-to-noise ratio (SNR) prevents the downloading of data for signals weaker than 25 dB/Hz. The data-download situation is improved by using *forward error correction codes* (FEC), and *block interleave* also covers for burst errors. FEC convolutional codes spread the information from one user data bit over many transmitted symbols. If some of these are lost, the data bit can be recovered from the others. However, a burst error may destroy all the relevant symbols. Interleaving, which transmits the symbols in a scrambled sequence, means that a single burst error cannot destroy all the symbols relevant to a single user data bit. The downside is that it adds latency to the message, to allow for the interleaving/de-interleaving process. However, on the Galileo signal with 1 s packets this is not a problem.

The 4 ms repetition rate is ideal because there is one symbol per code epoch. When the code is synchronized, we know that we will not hit a data bit edge because these always occur at the start of a code sequence.

The signal is the product of carrier, spreading code, BOC, and data. Traditionally, the RF hardware removes the carrier, the correlators remove the BOC(1,1) code, leaving the data and the residual Doppler to be removed/measured by a processor. With the BOC(1,1) codes, the BOC component should have been considered part of the spreading code for tracking and positioning; but it is equally viable to consider it part of the carrier during the acquisition phase, and remove it prior to the empirical correlation of acquisition.

The ACF of a BOC(1,1) code has three peaks, not just one, so care must be taken to ensure that the correct one has been found.

3.2   Galileo L1 OS Signal

In the following we describe and combine all elements necessary to generate the Galileo L1 OS signal.

The transmitted bandwidth is $40 \times 1.023\,\text{MHz} = 40.92\,\text{MHz}$. The minimum received power for the L1 OS signal is $-157\,\text{dBW}$ for elevation angles between $10°$ and $90°$. The chip length of the ranging code is

$$T_{c,\text{L1-B}} = T_{c,\text{L1-C}} = 1/1.023\,\text{Mchip/s} = 977.5\,\text{ns}. \tag{3.1}$$

The actual chips for the individual satellites are likely to be generated as a truncated Gold code. Higher chipping rates provide better accuracy. Longer codes reduce cross correlation to more acceptable levels, although acquisition time is longer.

The corresponding *ranging code rates* are

$$R_{c,\text{L1-A}} = 2.5 \times 1.023\,\text{Mchip/s},$$
$$R_{c,\text{L1-B}} = 1/T_{c,\text{L1-B}} = 1.023\,\text{Mchip/s},$$
$$R_{c,\text{L1-C}} = 1/T_{c,\text{L1-C}} = 1.023\,\text{Mchip/s},$$

and subcarrier rates

$$R_{sc,\text{L1-B}} = R_{sc,\text{L1-C}} = 1.023\,\text{MHz}.$$

Channel C uses both a primary and a secondary code of length $N_P = 4092$ chips and $N_S = 25$ chips, respectively. The *primary code* is a truncated Gold sequence, so when the number of 4092 chips is reached, the register is reset to its initial state. We remember a Gold sequence is obtained as modulo-2 addition of the output from two shift registers. The initial state for the first register consists of mere ones while the second register depends on the specific subcarrier and satellite.

The *secondary code* modulates 25 specific repetitions of the primary code. For each subcarrier all satellites transmit the same secondary code: the octal sequence 34,012,662. The resulting code length is 4092×25. In Galileo lingo the final code is called a *tiered code*.

Let the primary code generator work with chip rate R_P. The secondary code generator has chip rate $R_S = R_P/N_P$, where N_P is the length in chips of the primary code. In all signal modulations the logical levels 1 and 0 are defined as signal levels -1 and 1 (polar non-return-to-zero representation).

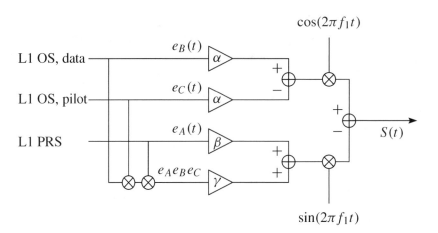

FIGURE 3.1. Galileo modulation scheme. The scheme is based on the modulation principle coherent adaptive subcarrier modulation (CASM).

Now we have information for defining the binary signal components for channels B and C. However, information on channel A is not available. Based on this information and the modulation scheme depicted in Figure 3.1, we will describe the Galileo signal L1 OS.

The signal component for channel B results from the modulo-2 addition of the navigation data stream d_{L1-B}, the PRN code sequence c_{L1-B}, and the B subcarrier sc_{L1-B}. The final component is called e_B. Likewise, the C channel results from the modulo-2 addition of the C channel PRN code sequence c_{L1-B} with the C channel subcarrier sc_{L1-C}. The component is e_C. The binary signal components are as follows:

$$e_A(t) = \text{not available}, \tag{3.2}$$

$$e_B(t) = \sum_{i=-\infty}^{+\infty} \left(c_{L1-B,(i \bmod 4\,092)} d_{L1-B,(i \bmod 4)} \text{rect}_{T_{c,L1-B}}(t - iT_{c,L1-B}) \right.$$
$$\left. \times \text{sign}\big(\sin(2\pi R_{c,L1-B}t)\big) \right), \tag{3.3}$$

$$e_C(t) = \sum_{i=-\infty}^{+\infty} \left(c_{L1-C,(i \bmod 4\,092)} \text{rect}_{T_{c,L1-C}}(t - iT_{c,L1-C}) \right.$$
$$\left. \times \text{sign}\big(\sin(2\pi R_{c,L1-C}t)\big) \right). \tag{3.4}$$

3.2.1   Signal Generation

Signal expressions are given for the power normalized complex envelope (i.e., baseband version) $s(t)$ of a modulated bandpass signal $S(t)$. Both are described in terms of its in-phase I and quadrature Q components by the following generic expressions:

$$S_{L1}(t) = \sqrt{2P_{L1}}\big(s_{L1-I}(t) \cos(2\pi f_{L1}t) - s_{L1-Q}(t) \sin(2\pi f_{L1}t)\big) \tag{3.5}$$

and normalized baseband signal

$$s(t) = s_{L1-I}(t) + j s_{L1-Q}(t). \tag{3.6}$$

3.2.2 Coherent Adaptive Subcarrier Modulation

The three channel signals $e_A(t)$, $e_B(t)$, and $e_C(t)$ of the L1 OS signal are multiplexed using coherent adaptive subcarrier modulation (CASM), (see Figure 3.1), which is a multichannel modulation scheme also known as tricode hexaphase modulation (or interplex modulation).

CASM is used to ensure that the signal transmitted from the satellite has a constant power envelope, i.e., the total transmitted power does not vary over time. Thus, the transmitted information is not contained in the signal amplitude and the transmitted signal amplitude becomes less critical. This is a very desirable property of the signal since it allows the use of efficient "class C"-like power amplifiers.

The L1 OS data and pilot signals are modulated onto the carrier in-phase component while the L1 PRS signal is modulated onto the quadrature component. The combined signal is

$$S(t) = \big(\alpha e_B(t) - \alpha e_C(t)\big) \cos(2\pi f_1 t) - \big(\beta e_A(t) + \gamma e_A(t) e_B(t) e_C(t)\big) \sin(2\pi f_1 t). \tag{3.7}$$

In this expression α, β, and γ are amplification factors that determine the distribution of useful power among the channels A, B, and C. We assume B and C have equal power.

For *given* relative signal powers we want to solve for these variables. So let us assume a relative signal power of 50% for A, and 25% for both B and C.

The first condition expresses that the norm of the I and the Q part of the signal S must be unity:

$$\sqrt{(\alpha - \alpha)^2 + (\beta + \gamma)^2} = 1.$$

The condition of equal power for both I and Q channels leads to

$$\alpha^2 + \alpha^2 = \beta^2.$$

Finally, the condition that the combined power equals one leads to

$$\alpha^2 + \alpha^2 + \beta^2 + \gamma^2 = 1.$$

In total, we have three equations in three unknowns:

$$\beta + \gamma = 1,$$
$$2\alpha^2 = \beta^2,$$
$$2\alpha^2 + \beta^2 + \gamma^2 = 1.$$

TABLE 3.1. Combination of three binary signals in CASM

L1 PRS	L1 OS D	L1 OS P	$\Re\big(S(t)\big)$	$\Im\big(S(t)\big)$	$|S(t)|$
1	1	1	0	1	1
1	1	−1	0.9428	0.3333	1
1	−1	1	−0.9428	0.3333	1
1	−1	−1	0	1	1
−1	1	1	0	−1	1
−1	1	−1	0.9428	−0.3333	1
−1	−1	1	−0.9428	−0.3333	1
−1	−1	−1	0	−1	1

The useful solution is

$$\alpha = \frac{\sqrt{2}}{3}, \qquad \beta = \frac{2}{3}, \qquad \gamma = \frac{1}{3}. \tag{3.8}$$

The given choice of relative signal powers defines the following signal:

$$S(t) = \frac{\sqrt{2}}{3}\Big(e_B(t) - e_C(t)\Big)\cos(2\pi f_1 t) - \frac{1}{3}\Big(2e_A(t) + e_A(t)e_B(t)e_C(t)\Big)\sin(2\pi f_1 t). \tag{3.9}$$

The product $e_A(t)e_B(t)e_C(t)$ is the intermodulation product L1 Int in CASM, which ensures the constant envelope property of the transmitted signal. The transmitted power is distributed as follows:

L1 OS, data $\alpha^2 = \left(\frac{\sqrt{2}}{3}\right)^2 = 22.22\%,$

L1 OS, pilot $\alpha^2 = \left(\frac{\sqrt{2}}{3}\right)^2 = 22.22\%,$

L1 P $\beta^2 = \left(\frac{2}{3}\right)^2 = 44.44\%,$

L1 Int $\gamma^2 = \left(\frac{1}{3}\right)^2 = 11.11\%.$

This means that only 88.88% of the total transmitted power is useful. The power offered for the L1 Int signal is wasted; apparently this waste is the price we must pay to obtain a constant envelope for the signal $S(t)$.

The L1 modulation scheme is specifically designed to allow independent processing of the L1 OS signal and the L1 PRS signal in the Galileo receiver. Only the quadrature component needs to be considered when the PRS signal is received. The in-phase part does not contain information about the PRS signal, as seen in Figure 3.1. Since the intermodulation product transmitted on the quadrature component does not carry any information, only the in-phase component is useful for reception of the L1 OS signals.

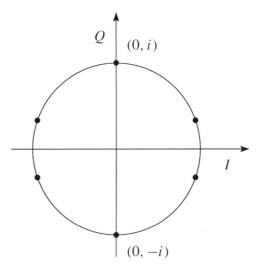

FIGURE 3.2. IQ plot of CASM output.

Table 3.1 shows the eight possible signal combinations of the three binary signals e_A, e_B, and e_C. Also shown are the real and imaginary parts of $S(t)$ (corresponding to the in-phase and quadrature signal components) as well as the norm of the IQ vector.

It is observed that the length of the IQ vector of $S(t)$ is always 1. This is also seen from the IQ plot in Figure 3.2 where the unit circle is included for reference.

From Table 3.1 it also appears that the three binary signals are combined to a representation with six possible phases, thus the name tricode hexaphase. It is seen from the table that the transmitted PRS code is equal to the sign of the quadrature component. It is also seen that the in-phase component can only assume three different values, so unambiguous mapping from received in-phase signal to transmitted signal is not possible. The table shows that the data channel is equal to the sign of the in-phase component when the data and pilot channels are not equal, and that the in-phase component is zero when the data channel is equal to the pilot channel. This means that all the transmitted power is concentrated in the quadrature channel whenever the data and pilot signals are equal.

3.2.3   Binary Offset Carrier Modulation

The Galileo signals and the planned modernized GPS signals inherit improved performance compared to the existing GPS signals. One of the improvements is the introduction of the binary offset carrier (BOC) modulation. BOC modulations offer two independent design parameters

- subcarrier frequency f_s in MHz, and

- spreading code rate f_c in Mchip/s.

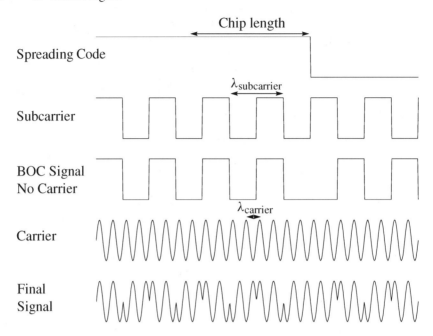

FIGURE 3.3. Spreading code, subcarrier, carrier, and signal as result of the BOC modulation principle. This figure does not show the navigation data.

These two parameters provide freedom to concentrate signal power within specific parts of the allocated band to reduce interference with the reception of other signals. Furthermore, the redundancy in the upper and lower sidebands of BOC modulations offers practical advantages in receiver processing for signal acquisition, code tracking, carrier tracking, and data demodulation; see Betz (2002).

Most Galileo signals come in pairs: a data signal and a data-free signal. They are aligned in phase and consequently have the same Doppler frequency.

A BOC(m, n) signal is created by modulating a sine wave carrier with the product of a PRN spreading code and a square wave subcarrier having each binary ± 1 values, see Figure 3.3. The parameter m stands for the ratio between the subcarrier frequency and the reference frequency $f_0 = 1.023$ MHz, and n stands for the ratio between the code rate and f_0. Thus, BOC$(10, 5)$ means a 10.23 MHz subcarrier frequency and a 5.115 MHz code rate.

The result of the subcarrier modulation is to split the classical BPSK spectrum in two symmetrical components with no remaining power on carrier frequency; see Martin et al. (2003).

The product is a symmetric split spectrum with two main lobes shifted from the carrier frequency by the amount equal to the subcarrier frequency; see Figure 3.4. We concentrate on BOC$(m, n) =$ BOC$(1, 1)$ as this is likely to be used by the L1 signal transmitted by Galileo.

The ACF of BOC signals has a profile with more peaks that may be tracked. For BOC signals it is important to make sure the channel is tracking the main peak of the correlation pattern. So additional correlators are needed for measuring the correlation profile at half a subcarrier phase from prompt correlator at either side.

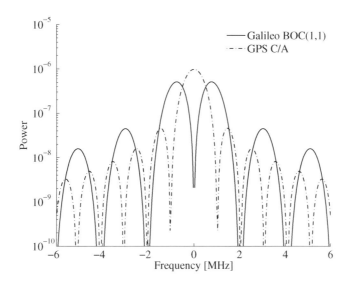

FIGURE 3.4. GPS C/A and Galileo BOC(1,1) sharing L1 spectrum. Center frequency is 1575.42 MHz.

If one of the output values of these so-called very early and very late correlators is higher than the punctual correlation, the channel is tracking a side peak and corrective action is taken.

According to Nunes & Sousa & Leitão (2004), the ACF for BOC(pn, n) with $p = 1, 2, \ldots$ and $k = \left\lceil \frac{2p|\tau|}{T_c} \right\rceil$ is given as

$$r(\tau) = \begin{cases} (-1)^{k+1}\left(\frac{1}{p}(-k^2 + 2kp + k - p) - (4p - 2k + 1)\frac{|\tau|}{T_c} \right), & \text{for } |\tau| \leq T_c, \\ 0, & \text{otherwise.} \end{cases}$$

(3.10)

This ACF is plotted in Figure 3.5. For $p = 1$ this is

$$r(\tau) = \begin{cases} (-1)^{k+1}\left(-k^2 + 3k - 1 - (5 - 2k)\frac{|\tau|}{T_c} \right), & \text{for } |\tau| \leq T_c, \\ 0, & \text{otherwise.} \end{cases}$$

The ACF for the BOC(n, n) signal with bandwidth b is given as [see Winkel (2000), Equation (2.68)]

$$r_{\text{BOC}}(\tau) = \sum_{k=-n+1}^{n-1}(n - |k|)\Big(2r_{\text{BL}}(\tau/T_c - 2k) - r_{\text{BL}}(\tau/T_c - 2k - 1) - r_{\text{BL}}(\tau/T_c - 2k + 1) \Big),$$

(3.11)

where

$$r_{\text{BL}}(t) = \tfrac{1}{\pi}(t + 1)\,\text{Si}\big(2\pi b(t + 1)\big) + \tfrac{1}{2\pi^2 b}\cos\big(2\pi b(t + 1)\big)$$
$$+ \tfrac{1}{\pi}(t - 1)\,\text{Si}\big(2\pi b(t - 1)\big) + \tfrac{1}{2\pi^2 b}\cos\big(2\pi b(t - 1)\big)$$
$$- \tfrac{2t}{\pi}\,\text{Si}\big(2\pi b t\big) - \tfrac{1}{\pi^2 b}\cos\big(2\pi b t\big) \quad (3.12)$$

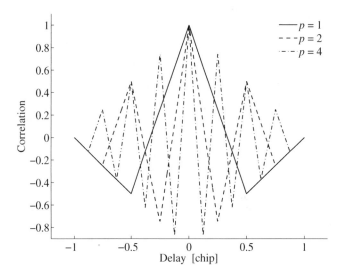

FIGURE 3.5. ACF for the BOC(pn, n) signal as function of delay τ and p.

and the sine integral is defined as

$$\text{Si}(x) = \int_0^x \frac{\sin(y)}{y}\, dy.$$

If we plot the function $r_{\text{BOC}}(n, n)$, we get a result similar to the one in Figure 3.5 for $n = 1, 2, 4$.

The BOC ACF profile results in a DLL discriminator curve that is a bit more complicated than that of GPS. Figure 3.6 shows the ideal band-unlimited correlation function for both a C/A code signal and a BOC(1,1) signal. Shown as well are early minus late discriminator curves for a chip spacing of 0.5 chip.

We observe various facts. Both discriminator curves are linear around the center of the ACF. In both cases the linear region extends from -0.25 to 0.25 chip code offset. The slope of the BOC discriminator in the linear region is three times the slope of the C/A discriminator. The C/A code discriminator output is used to adjust the code NCO to align the code phase better with the incoming signal; this adjustment will succeed for tracking errors less than 1.25 chips. The C/A dis-

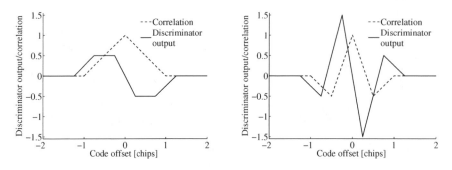

FIGURE 3.6. C/A code and BOC(1,1) ACF and early minus late discriminator curves.

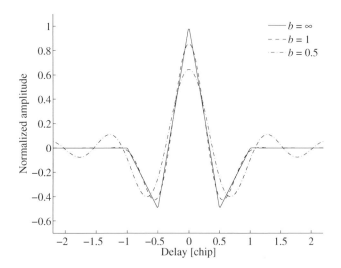

FIGURE 3.7. ACF for bandlimited BOC(1,1) signal. The normalized bandlimit is $b = 0.5$, 1, and ∞. The function for $b = \infty$ is identical to BOC(1,1) in Figure 3.5.

criminator is stable in the entire region where the discriminator curve is non-zero and the DLL will converge. The BOC discriminator has stable regions next to the linear region as well, but tracking errors in the outer regions (absolute errors less than 1.25 and greater than 0.625 chip) will cause the DLL to diverge and lose lock.

Figure 3.4 is a nice illustration of the fact that the GPS C/A code and the Galileo BOC(1,1) modulation share the L1 spectrum. Still it is possible to separate the two signals.

According to Betz (2002), the power spectral density of the BOC($f_s/f_o, f_c/f_o$) centered at the origin is

$$S(\omega) = f_c \left(\frac{\tan\left(\frac{\pi\omega}{2f_s}\right)\sin\left(\frac{\pi\omega}{f_c}\right)}{\pi\omega} \right)^2, \qquad \frac{2f_s}{f_c} = n \quad \text{even,} \qquad (3.13)$$

$$S(\omega) = f_c \left(\frac{\tan\left(\frac{\pi\omega}{2f_s}\right)\cos\left(\frac{\pi\omega}{f_c}\right)}{\pi\omega} \right)^2, \qquad \frac{2f_s}{f_c} = n \quad \text{odd.} \qquad (3.14)$$

The number of negative and positive peaks is $2n - 1$ in the ACF separated in delay by $T_s = 1/2p$.

For the Galileo BOC(1,1) data-carrying signal, the code length is 4092 chips, 1.023 MHz band frequency (4 ms period time), 250 Hz symbol rate (1 code period per symbol), $\sim 85\%$ of signal power within ~ 4 MHz bandwidth.

Figure 3.7 shows the ACF for bandlimited BOC(1,1) signals. For limited bandwidth the peak value is less than one; this reflects the fact that not all power is available in the signal. Part of the power is blocked by the bandlimiting. For $b = 1$ the bandlimiting results in a slight rounding off at the edges of the ACF. For $b = 0.5$ the frequencies lower than twice the square wave frequency are stopped by the filter. This results in oscillations outside the chip length region. This could lead to undesirable side-lobe effects in case of multipath.

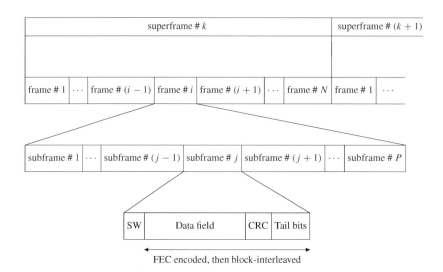

FIGURE 3.8. Navigation message structure. CRC is computed over the data field.

3.3   Message Structure

This section outlines the Galileo message structure.

3.3.1   Frames and Pages

The message is composed of *frames*, see Figure 3.8. The frame is composed of several subframes, and each subframe again is composed of several pages. The *page* is the basic structure for the navigation message and contains the following fields:

- a synchronization word (SW),
- a data field,
- a cyclic redundancy check (CRC) bits for error detection, and
- tail bits for the forward error correction (FEC) encoder containing all zeros.

CRC and data encoding are used to provide better signal and data integrity. For L1 OS the synchronization word is a fixed 10-bit sequence.

All data are encoded using the following bit and byte ordering:

- for numbering, the most significant bit/byte is numbered as bit/byte 0,
- for bit/byte ordering, the most significant bit/byte is transmitted first.

In Figure 3.9 the most significant bit is placed left, the less significant bit is placed to the right, the most significant items at the top, and the less significant items at the bottom.

3.3.2   Cyclic Redundancy Check

The CRC algorithm accepts a binary data frame, corresponding to a polynomial M, and appends a checksum of r bits, corresponding to a polynomial C.

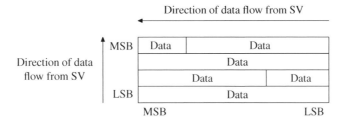

FIGURE 3.9. Ordering principle for data.

The concatenation of the input frame and the checksum then corresponds to the polynomial $T = Mx^r + C$ since multiplying by x^r corresponds to shifting the input frame r bits to the left. The algorithm chooses the checksum C such that T is divisible by a predefined polynomial P of degree r, called the *generator polynomial*.

The algorithm divides Mx^r by P and sets the checksum equal to the binary vector corresponding to the remainder. That is, if $Mx^r = QP + R$, where R is a polynomial of degree less than r, then $C = R$ and the checksum is the binary vector corresponding to R. If necessary, the algorithm precedes zeros to the checksum so that it has length r.

The MATLAB CRC generator does the following:

1. Left-shift the input frame by r bits and divide the corresponding polynomial by P.

2. Set the checksum equal to the binary vector of length r corresponding to the remainder from step 1.

3. Append the checksum to the input data frame. The result is the output frame.

The CRC algorithm uses binary vectors to represent binary polynomials, in descending order of powers. For example, the vector [1 1 0 1] represents the polynomial $x^3 + x^2 + 1$.

Example 3.1 Suppose the input frame is $[1\ 1\ 0\ 0\ 1\ 1\ 0]^T$, corresponding to the polynomial $M = x^6 + x^5 + x^2 + x$, and the generator polynomial is $P = x^3 + x^2 + 1$, of degree $r = 3$. By polynomial division $Mx^3 = (x^6 + x^3 + x)P + x$. Remember that any binary number added to itself in a modulo-2 field yields zero. The remainder is $R = x$ so that the checksum is then $[0\ 1\ 0]^T$. Note that an extra 0 is added on the left to make the checksum have length 3.

3.3.3   Forward Error Correction and Block Interleaving

The starting point is a digital information source (transmitter) that sends a data sequence comprising k bits of data to an encoder. Employing *forward error-correction* coding, the encoder inserts redundant bits, thereby outputting a longer sequence of n code bits called a *codeword*. At the receiving end, codewords are used by a suitable decoder to extract the original data sequence.

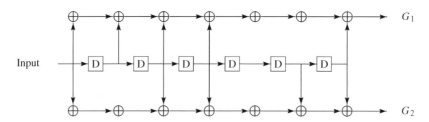

FIGURE 3.10. Viterbi convolutional coding scheme.

In general, codes are designated with the notation (n, k) according to the number of n output code bits and k input data bits. The ratio k/n is called the *rate of the code* and is a measure of the fraction of information contained in each code bit.

Figure 3.10 describes the Viterbi convolutional coding scheme used for Galileo. The Viterbi convolutional coding of all data channels is characterized by the following values: coding rate $= 1/2$, constraint length 7, generator polynomials $G_1 = 171$ (octal), $G_2 = 133$ (octal), and encoding sequence G_1 and then G_2. That is, there are taps at seven points along the encoder shift register, and it generates two encoded channel symbols for each input data bit. Note that $G_1 = 171$ (octal) is 001111001 in binary and $G_2 = 133$ (octal) is 001011011 in binary.

We observe that for G_1 the taps 1, 4, 5, 6, and 7, numbered from right to left, are connected to modulo-2 adders and for G_2 the taps 1, 2, 4, 5, and 7 are connected to modulo-2 adders. The error-correcting power is related to the constraint length, increasing with longer lengths of shift register.

After the Viterbi convolutional encoding block follows an interleaving procedure. The purpose of *interleaving* is to increase the efficiency of the convolutional coding by spreading the burst errors in order to improve the correction capacity of the specific convolutional decoding algorithm applied in the receiver.

Normally interleaving is implemented by using a two-dimensional buffer array. The data enter the buffer array row by row and are then read out column by column. The result of the interleaving process is ideal if a burst of errors in the communication channel after de-interleaving becomes spaced single-symbol errors, which are easier to correct than consecutive ones.

Example 3.2 Below we illustrate how an input data stream is read, column by column, into an array and then read out, row by row, to achieve the bit interleaving operation:

– Data input

$$x_{16}\, x_{15}\, x_{14}\, x_{13}\, x_{12}\, x_{11}\, x_{10}\, x_9\, x_8\, x_7\, x_6\, x_5\, x_4\, x_3\, x_2\, x_1$$

– Interleaving array

x_1	x_5	x_9	x_{13}
x_2	x_6	x_{10}	x_{14}
x_3	x_7	x_{11}	x_{15}
x_4	x_8	x_{12}	x_{16}

TABLE 3.2. Galileo ephemeris parameters

Parameter	No. of Bits	Scale Factor	Unit
μ_0	32	2^{-31}	semicircle
Δn	16	2^{-43}	semicircle/s
e	32	2^{-33}	dimensionless
$\sqrt{a}$	32	2^{-19}	$m^{1/2}$
Ω_0	32	2^{-31}	semicircle
i_0	32	2^{-31}	semicircle
ω	32	2^{-31}	semicircle
$\dot{\Omega}$	24	2^{-43}	semicircle/s
$\dot{i}$	14	2^{-43}	semicircle/s
C_{uc}	16	2^{-29}	radian
C_{us}	16	2^{-29}	radian
C_{rs}	16	2^{-6}	m
C_{rc}	16	2^{-6}	m
C_{ic}	16	2^{-29}	radian
C_{is}	16	2^{-29}	radian
t_{oe}	14	60	s
IOD_{nav}	9	—	—

– Interleaved data output for transmission

$$x_{16} \; x_{12} \; x_8 \; x_4 \; x_{15} \; x_{11} \; x_7 \; x_3 \; x_{14} \; x_{10} \; x_6 \; x_2 \; x_{13} \; x_9 \; x_5 \; x_1$$

This block interleaving has interleaving depth 4. In this simple example three consecutive errors in the transmitted data, let us say x_1, x_5, x_9, are translated into three isolated, single errors in the de-interleaved data.

3.4 Message Contents

The freely accessible navigation message on L1 OS contains all parameters necessary to compute the position of a Galileo satellite, clock correction parameters, parameters for conversion of Galileo System Time (GST) to UTC and GPS Time (GPST), and service parameters including an almanac.

A Galileo ephemeris contains 17 parameters as defined in Table 3.2. The unit semicircle is converted to radian by multiplication with π.

A Galileo ephemeris is valid for four hours. Every three hours a new ephemeris is uploaded. Hence an overlap of one hour occurs. And four ephemerides cover a 12-hour orbit prediction.

The geocentric coordinates (X^k, Y^k, Z^k) of satellite k (the superscript k denotes satellite k, and not the power k) at time t_j are given as

$$\begin{bmatrix} X^k(t_j) \\ Y^k(t_j) \\ Z^k(t_j) \end{bmatrix} = R_3(-\Omega_j^k)R_1(-i_j^k)R_3(-\omega_j^k)\begin{bmatrix} r_j^k \cos f_j^k \\ r_j^k \sin f_j^k \\ 0 \end{bmatrix}. \qquad (3.15)$$

For a definition of the parameters Ω, i, ω, and f, see Figure 8.5. The quantities $r_j^k = \|r(t_j)\|$, a, e, and E are evaluated for t according to the following procedure:

Time elapsed since t_{oe} $\qquad\qquad t_j = t - t_{oe}$

Mean anomaly at time t_j $\qquad\qquad \mu_j = \mu_0 + \left(\sqrt{GM/a^3} + \Delta n\right)t_j$

$$GM = 3.986\,005 \cdot 10^{14}\,\text{m}^3/\text{s}^2$$

Iterative solution for E_j $\qquad\qquad E_j = \mu_j + e \sin E_j$

True anomaly $\qquad\qquad f_j = \arctan \dfrac{\sqrt{1 - e^2}\sin E_j}{\cos E_j - e}$

Longitude for ascending node $\qquad \Omega_j = \Omega_0 + (\dot{\Omega} - \omega_e)t_j - \omega_e t_{oe}$

$$\omega_e = 7.292\,115\,147 \cdot 10^{-5}\,\text{rad/s}$$

Argument of perigee $\quad \omega_j = \omega + f_j + C_{\omega c}\cos 2(\omega + f_j) + C_{\omega s}\sin 2(\omega + f_j)$

Radial distance $\qquad\quad r_j = a(1 - e\cos E_j) + C_{rc}\cos 2(\omega + f_j)$

$$+ C_{rs}\sin 2(\omega + f_j)$$

Inclination $\qquad\qquad i_j = i_0 + \dot{i}t_j + C_{ic}\cos 2(\omega + f_j) + C_{is}\sin 2(\omega + f_j).$

As usual the mean Earth rotation rate is denoted ω_e. This algorithm is similar to the one for GPS and is coded as the M-file satpos. The function computes the position of any Galileo satellite at any time. It is fundamental to every position calculation.

3.4.1   Time and Clock Correction Parameters

As for GPS, Galileo has its own system time, called Galileo System Time (GST). Its starting epoch still has to be determined. GST consists of two parts: week number, WN, and time of week, TOW. The WN covers 4096 weeks and is then reset to zero. A week has 604,800 s and is reset at midnight between Saturday and Sunday. Hence GST is described as a 32-bit binary number split into the two parts just mentioned. Table 3.3 shows these parameters.

Let a signal be transmitted at time $t^{k,\text{Gal}}$ from satellite k, and let the same signal be received at time t_i^{Gal} at receiver i. Then the travel time is

$$\tau_i^k = t_i^{\text{Gal}} - t^{k,\text{Gal}}. \qquad (3.16)$$

TABLE 3.3. Galileo system time parameters

Parameter	No. of bits	Scale factor	Unit
WN	12	—	week
TOW	20	1	s

TABLE 3.4. Galileo clock correction parameters

Parameter	No. of bits	Scale factor	Unit
t_{oc}	14	60	s
a_o	28	2^{-33}	s
a_1	18	2^{-45}	s/s
a_2	12	2^{-65}	s/s^2

Knowing the travel time τ_i^k this quantity can be converted to the so-called pseudorange P_i^k by multiplication with the speed of light c:

$$P_i^k = c\tau_i^k. \tag{3.17}$$

However, clocks do not work perfectly. So we introduce the receiver clock offset dt_i and the satellite clock offset dt^k:

$$t_i = t^{\mathrm{Gal}} + dt_i$$
$$t^k = (t - \tau_i^k)^{\mathrm{Gal}} + dt^k.$$

The receiver clock offset has to be estimated from the observed pseudoranges while the satellite clock offset can be computed from

$$dt^k = a_0 + a_1(t - t_{0c}) + a_2(t - t_{0c})^2, \tag{3.18}$$

where t is the transmit time. The constants a_0, a_1, and a_2 are parameters transmitted according to Table 3.4.

The *basic computational equations for time* are

$$t^{k,\mathrm{Gal}} = t_i^{\mathrm{Gal}} - P_i^k/c \tag{3.19}$$
$$(t - \tau_i^k)^{\mathrm{Gal}} = t^k - \left(a_0 + a_1(t - t_{0c}) + a_2(t - t_{0c})^2\right). \tag{3.20}$$

Additionally, a signal in space accuracy (SISA) parameter is planned. It is encoded as 8 bits.

An *ionospheric correction service* is planned. An effective ionization-level parameter is computed from three broadcast coefficients.

TABLE 3.5. GST to UTC conversion

Parameter	No. of bits	Scale factor	Unit
A_0	32	2^{-30}	s
A_1	24	2^{-50}	s/s
Δt_{LS}	8	1	s
t_{0t}	8	3600	s
WN_t	8	1	week
WN_{LSF}	8	1	week
DN	3	$1 \ldots 7$	day
Δt_{LSF}	8	1	s

3.4.2 Conversion of GST to UTC and GPST

Compared to the present GPS, Galileo offers some advantages for the timing community. For example, data for real-time estimation of Universal Time Coordinated (UTC) are available. Likewise for the difference between GST and GPST. However, in case the user disposes over a combined GPS and Galileo receiver, it is likely that an estimate based on Equation (8.45) turns out to be more accurate.

The relation between GST and UTC is established via the time scale Temps Atomique International (TAI). UTC and TAI differ by an integer number of seconds. On January 1, 2003, the difference was

$$TAI - UTC_{2003} = +32\,s.$$

UTC is a uniform time scale, and it tries to follow variations in the Earth's rotation rate; this is accommodated for by introducing leap seconds in the UTC. Consequently, this changes the difference between UTC and GST in steps of 1 s (see Table 3.5).

Let the estimated epoch time in GST, relative to the start of the week, be denoted by t_E. Let A_0 denote the offset between GST and TAI at the time t_E. The time derivative of A_0 is called A_1. Let the difference between TAI and UTC be Δt_{LS}, and the validity time t_{0t} for the UTC offset parameters.

Leap seconds are always introduced on January 1 or/and July 1. The day number in the week in which the leap second is introduced is called DN. Days are counted from 1 to 7 (Sunday is 1) and is rounded to an integer.

The week number, modulo 256, in which DN falls is denoted WN_{LSF}. Finally, the offset due to the introduction of a leap second at WN_{LSF} and DN is called Δt_{LSF}.

The following equations are in unit of s. We start by introducing the correction

$$\Delta t_{UTC} = \Delta t_{LS} + A_0 + A_1\big(t_E - t_{0t} + 604{,}800(WN - WN_t)\big). \qquad (3.21)$$

TABLE 3.6. GST to GPST conversion

Parameter	No. of bits	Scale factor	Unit
A_{0G}	16	2^{-35}	s
A_{1G}	12	2^{-51}	s/s
t_{0G}	8	3600	s

We need to distinguish among three different cases:

– When $t_E > \text{WN}_{\text{LSF}}$ and $t_E > \text{DN}$ and $\text{DN} + \frac{3}{4} < t_E$ and $t_E < \text{DN} + \frac{5}{4}$, then

$$t_{\text{UTC}} = \text{mod}(t_E - \Delta t_{\text{UTC}}, 86{,}400), \qquad (3.22)$$

– when $\text{DN} + \frac{3}{4} < t_E < \text{DN} + \frac{5}{4}$, then

$$W = \text{mod}(t_E - \Delta t_{\text{UTC}} - 43{,}200, 86{,}400) + 43{,}200 \qquad (3.23)$$

$$t_{\text{UTC}} = \text{mod}(W, 86{,}400 + \Delta t_{\text{LSF}} - \Delta t_{\text{LS}}), \qquad (3.24)$$

– when $\text{WN}_{\text{LSF}} < t_E$ and $\text{DN} > t_E$, then

$$t_{\text{UTC}} = \text{mod}(t_E - \Delta t_{\text{UTC}}, 86{,}400). \qquad (3.25)$$

The *difference between GST and GPST* is determined as follows. Let t_{Gal} be GST estimated by the user receiver; then the offset between GST and GPST at time t_{Gal} is

$$\Delta t_{\text{systems}} = A_{0G} + A_{1G}(t_{\text{Gal}} - t_{0G}). \qquad (3.26)$$

Table 3.6 describes the parameters concerned.

3.4.3   Service Parameters

The satellite identification SV_{ID} is a number between 1 and 128. A parameter, issue of data (IOD), identifies the set of data. This allows a receiver to compare batches of data received from different satellites. IOD is transmitted in each page of ephemeris and clock correction (9 bits) and almanac (2 bits).

Signal and data health status referring to the transmitting satellite is planned as well.

The six Keplerian elements $(a, e, \omega, \Omega, i, \mu)$ of any active satellite are contained in the *almanac*. The elements are given with less precision than the ephemeris. Clock correction parameters are given for computation of satellite clock offset

$$dk^k = a_{f0} + a_{f1}(t - t_{0a}). \qquad (3.27)$$

The almanac reference time t_{0a} refers to the almanac reference week WN_a.

TABLE 3.7. Almanac parameters

Parameter	No. of Bits	Scale Factor	Unit
SV_{ID}	7	1	dimensionless
$\sqrt{a}$	24	2^{-11}	$m^{1/2}$
e	16	2^{-21}	dimensionless
i_0 (relative to $56°$)	16	2^{-19}	semicircle
Ω_0	24	2^{-23}	semicircle
$\dot{\Omega}$	16	2^{-38}	semicircle/s
ω	24	2^{-23}	semicircle
μ	24	2^{-23}	semicircle
a_{f0}	15	2^{-20}	s
a_{f1}	11	2^{-38}	s/s
SV_{SHS}	5	—	dimensionless
SV_{DHS}	3	—	dimensionless
$Data_{ID}$	2	—	dimensionless
IODA	2	—	dimensionless
t_{0a}	8	4096	s
WN_a	8	1	week

Two parameters tell the satellite's signal component health SV_{SHS} and the satellite's navigation data health SV_{DHS}. In the almanac the applicable navigation data structure for each satellite is defined by $Data_{ID}$. The IODA identifies an almanac batch unambiguously. The update rate being slow, two bits are sufficient. All parameters are described in Table 3.7.

3.5   The Received L1 OS Signal

Let the total received power be P, the transmission delay (traveling time) be τ, the carrier frequency offset be Δf (Doppler), and the received phase be θ. Then the *received L1 OS signal* can be written as

$$\varphi(t) = 0.89 \frac{\sqrt{2P}}{3} \Big(s_d(t - \tau) - s_p(t - \tau) \Big) \cos\big(2\pi(f - \Delta f)(t - \tau) + \theta\big). \quad (3.28)$$

The data channel and the pilot channel are denoted by d and p, respectively. The coefficients s_d and s_p are products of code sequences and subcarriers with sine phasing.

From the observation $\varphi(t)$ we want to estimate τ, Δf, and θ. The first step is to find global approximate values of τ and Δf, which is called *signal acquisition*. The second step is a local search for τ, Δf, and possibly θ. If θ is estimated, the search is called *coherent signal tracking*. If the carrier phase θ is ignored, the search is called *noncoherent signal tracking*.

The purpose of code tracking is to estimate the travel time τ and is done by means of a *delay lock loop* (DLL). For a coherent DLL we have $\theta = 0$.

To demodulate the navigation data, a carrier wave replica must be generated. To track a carrier wave signal, a *phase lock loop* (PLL) often is used.

A final remark.   This chapter exposed most of the material relevant to code the L1 OS Galileo signal. However, GPS is also under continuous development— a very fortunate situation for the user. The planned civilian L5 GPS signal is described in ICD-GPS-705 (2002). Comparing the present chapter, which is based on Anonymous (2005), with ICD-GPS-200 (1991) you realize that the European and American satellite navigation communities are each using their own lingos.

4

GNSS Antennas and Front Ends

4.1 Background

Although the focus of this text is on the algorithms for software signal processing of the Global Navigation Satellite System (GNSS) signals, it is important to consider the source of that data stream to be processed. Since "software" signal processing is stated, it implies an input digital data stream. Thus, the purpose of this chapter is to provide some insight into how the satellite signals propagating through space result in this digital data stream. This is done, of course, via a GNSS antenna/front end. There are numerous books completely devoted to the topics of antenna and others to front-end design; see Balanis (1996) and Tsui (2000). The purpose of this chapter is to illustrate functional designs for GNSS, discuss the tradeoffs associated with different designs, and provide a basic understanding of the source of the digital data to be processed. The focus is on the narrowband GNSS L1 signals, primarily the Global Positioning System (GPS) Coarse/Acquisition (C/A) code, but references are made to the Galileo BOC(1,1) code where appropriate. At the end of the chapter, multiple-band GNSS front ends are introduced.

The process begins with the GNSS signal, propagating through space, which is incident on a user's GNSS antenna. This, in turn, induces a voltage within the element. That voltage is extremely weak, corresponding to a guaranteed signal power of -160 dBW in the case of the Global Positioning System (GPS) [see ICD-GPS-200 (1991)] and has a carrier frequency of 1575.42 MHz. Considering a bandwidth of 2 MHz (the approximate null-to-null bandwidth the GPS C/A code signal), the received GPS signal power is actually below that of the thermal noise floor, as defined by Equation (4.1) with a simplified illustration in Figure 4.1.

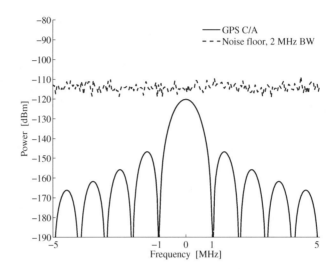

FIGURE 4.1. Frequency domain depiction of the GPS signal and thermal noise power. Remember that $30\,\mathrm{dBm} = 1\,\mathrm{dBW}$. Center frequency $1575.42\,\mathrm{MHz}$.

Let the Boltzmann's constant be denoted by $k = 1.38 \cdot 10^{-23}\,\mathrm{J/^\circ K}$, the absolute temperature by t in $^\circ\mathrm{K}$, and the equivalent noise bandwidth by B in Hz, then

$$P_{\text{Thermal Noise}} = ktB. \tag{4.1}$$

For the GPS C/A code signal $P_{\text{Thermal Noise}}$ can be approximated by $1.38 \cdot 10^{-23} \times 290 \times 2 \cdot 10^6 = 8.004 \cdot 10^{-15}$ or more conveniently expressed in dB as $10 \times \log_{10}(8.004 \cdot 10^{-15}) = -140.97\,\mathrm{dBW} = -110.97\,\mathrm{dBm}$.

This is quite unique in the field of radio transmission. For example, if you connected a traditional GPS antenna to a spectrum analyzer and searched for the presence of the GPS signal, then any such characteristics of the signal would be hidden as the observation would be dominated by the thermal noise. This is a feature of the code division multiple access (CDMA) spread spectrum signal and requires the appropriate signal processing to acquire and process the signal. This also implies that the design of the front end is based more on the level of the thermal noise rather than the received L1 band navigation signal. Thus, the voltage induced within the GNSS antenna element results from the thermal noise, which dominates, as well as the GNSS signals from the satellites in view. Given that that L1 GNSS band is a designated Aeronautical Radio Navigation Service frequency band, no other signals should be present within the frequency span.

The analog voltage that results from the incident GPS signal and thermal noise remains much too weak and at too high a frequency for most analog-to-digital converters (ADCs) to operate. In order to overcome this, the front end will utilize a combination of amplifier(s), mixer(s), filter(s), and its own oscillator to condition the incident voltage on the antenna to the resulting digital samples.

A fully functional GNSS L1 front end is depicted in Figure 4.2. In the coming sections, the function of each of the elements within the figure will be discussed using this implementation as a case study, Gromov et al. (2000), pages 447–457.

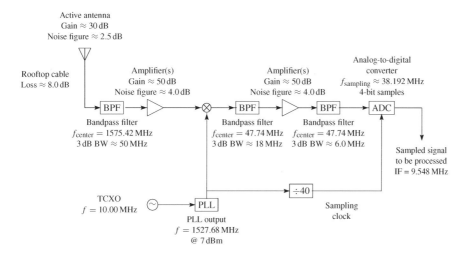

FIGURE 4.2. GNSS L1 front end.

4.2   GNSS L1 Front-End Components

4.2.1   GNSS Antenna

The antenna is typically not considered part of the front-end design, but since it is the first component in the signal path and dictates elements that follow, it is important to summarize when describing the GNSS front end. There are numerous texts on antenna theory and design, e.g., Balanis (1996), Straw (2003). Also the trade publication *GPS World*, over the past four years, has published a GPS Antenna Survey that lists all GPS antennas and their features. All of these are excellent references for additional information on GNSS antennas.

As is the case with most of the components associated with analog signal conditioning, there is an extensive set of parameters associated with an antenna that describe its performance. Three fundamental parameters to be discussed here are the *frequency/bandwidth*, *polarization*, and *gain pattern*.

The antenna will be designed to induce a voltage from radio waves propagating at the GNSS L1 frequency or 1575.42 MHz. In addition, the design should accommodate the appropriate bandwidth of the desired signal. This is usually specified using two additional antenna parameters: the *Voltage Standing Wave Ratio* (VSWR) and *impedance*. Practically all GNSS front-end components utilize an impedance of 50 Ω, which is typical for a majority of radio frequency design. VSWR is a measure of impedance mismatch or the measure of how much of the incident power will be absorbed and how much will be reflected. And, of course, this is a function of frequency. The VSWR is typically on the order of 2.0:1, which equates to 90% power absorption across the bandwidth of desired frequencies.

Polarization refers to the orientation of the electric field from the radio frequency transmission. Received GNSS signals are right-hand circularly polarized (RHCP), and the antenna should be designed as such. The decision to employ RHCP for GNSS was definitely not arbitrary. One of the most difficult error sources to mitigate for GNSS is multipath. When the GNSS signal is reflected off an object, an undesirable situation for a system attempting to measure time-of-flight, the polarization will flip to left-hand circular polarization (LHCP). An RHCP antenna is quite effective in suppressing the LHCP reflection and minimizing this error source. Of course, a second reflection will reestablish the RHCP polarization, but the signal power is also likely diminished as a result of the multiple reflections. Thus, the polarization of the GNSS antenna provides a significant level of suppression from erroneous multipath reflection.

The antenna pattern describes the directivity of the antenna. The most basic idea for the antenna pattern would be one that receives signals equally from all directions—this is known as an isotropic antenna. However, such a uniform gain pattern does not make sense for GNSS. Since the signal source, GNSS satellites, are overhead for most applications the preferred antenna pattern would be hemispherical, designed to receive signal from only positive elevation angles from all azimuth directions. Given the problem of multipath and that most multipath rays arrive from low elevation angles, the antenna pattern could be crafted such that it was designed to receive signals only above 10°–20° elevation. Such an approach is definitely bound to further reduce multipath reflections, but as a consequence, the low elevation satellite signals would also be neglected, decreasing the availability of satellite measurements. A promising research area within GNSS antennas is that of antenna arrays, or a combination of individual antenna elements combined in such a way to shape distinct antenna pattern beams and nulls. Such an implementation should provide significant performance enhancement for GNSS.

Probably two of the most popular GNSS L1 antenna implementations are the patch and helix approaches but others also exist. These refer to the actual construction of the antenna element itself. Yet the parameters above should provide a measure of comparable performance between antennas.

The last topic in regard to antennas refers to the choice of an *active* or *passive* antenna. An antenna will often be integrated with other front-end components that improve their performance or are necessitated by the environment in which the antennas will operate.

One important parameter in front-end design is the overall *noise figure F_n* of the system. This parameter quantifies the noise added as a result of the analog signal conditioning. Of course, any additive noise or decrease in signal-to-noise ratio (SNR) is undesirable and should be minimized.

Denoting the resulting system noise figure by F_{system}, the noise figure F_n of the nth element in cascade, and the gain of the nth element in cascade G_n the formula for noise figure is

$$F_{\text{system}} = F_1 + \frac{F_2 - 1}{G_1} + \frac{F_3 - 1}{G_1 G_2} + \frac{F_4 - 1}{G_1 G_2 G_3} + \cdots + \frac{F_n - 1}{G_1 G_2 G_3 \cdots G_{n-1}}. \quad (4.2)$$

What this equation indicates is that the first element in the RF cascade dominates the resulting noise figure for the system. This indicates that all passive components (cables and filters) prior to the first amplifier will have a negative impact on the noise figure. Likewise, components that follow a high gain amplifier in the cascade will have a minimal effect on the overall noise figure.

For example, consider working with a GNSS receiver in a laboratory environment. The optimal position for the GNSS antenna will be on the rooftop, clear of any obstructions. In most cases, this will require a lengthy cable run to the GNSS receiver with its self-contained front end. This RF cable from the antenna to the front end will be the first component within the cascade of components. Since all RF cables have some degree of attenuation, or noise figure, and no gain, the system noise figure will be severely degraded. This can be improved if an amplifier can be incorporated within the antenna itself prior to the long cable. This implementation is the norm in many GNSS antennas, and such a design is known as an active antenna and is characterized by the gain of the amplifier.

This *active antenna* approach complicates things slightly as the antenna itself is now considered an active element and requires power for the internal amplifier. This is accomplished in most cases using a bias-tee. The bias-tee component has three ports: RF, RF+DC, and DC. This component injects DC power onto the antenna cable from the front end to power the amplifier within the antenna. Thus, the antenna cable is utilized to pass the GNSS signal from the antenna to the bulk of the analog signal conditioning and then also to provide a DC voltage from the analog signal conditioning to the amplifier within the antenna. This is the approach outlined in Figure 4.2.

A *passive antenna* is practical in those designs that have the antenna in close proximity to the analog signal conditioning and, in particular, the first amplifier. This is commonly the case in the handheld GNSS receivers or for configurations employing expensive low-loss RF cables.

4.2.2   Filter

The first component within the RF path is a filter. A filter is a frequency selective device that allows only certain frequencies to pass and attenuates others.

The treatment of the filter as well as the following individual components will be kept terse. It is expected the reader has a basic background in signal processing; this will allow the focus to be on the overall GNSS front-end design.

This first filter in Figure 4.2 is a bandpass filter, as opposed to a lowpass or highpass filter, and its purpose is to provide additional frequency selectivity. Ideally, the antenna would only induce voltages for precisely the frequency band of interest. However, the antenna, like practical filters, is not ideal. The ideal component would pass a range of frequencies and completely eliminate those frequency components outside that range. Unfortunately, such a filter does not exist, and the transition between those frequencies that are passed and removed is a gradual transition. Further, even signals at frequencies within the passband typically experience some level of attenuation.

Typical antennas have fairly poor frequency selectivity. When this is considered, along with the received signal power levels (and the amplification that will be required), it is important to try to eliminate any high-power, out-of-band signal

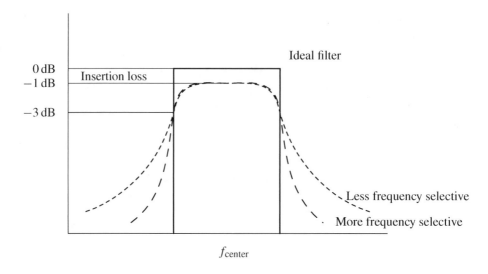

FIGURE 4.3. Comparison of bandpass filters.

sources that could enter the front end and saturate later-stage components. For this reason, a bandpass filter typically is the first component following the antenna. The other filters within Figure 4.2 serve specific roles, as will be discussed in the following sections, but all are used to perform the primary role of a filter: pass selective frequencies and attenuate others.

Filters can be characterized by their *insertion loss*, or the attenuation of the desired frequency components. Ideally, there would be no insertion loss, but alas that is not the case for practical components, and the lower this value the better. Note that this filter insertion loss will result in a system noise figure penalty when it is placed prior to the first amplifier. Yet this is still very often the case to minimize any issues from adjacent frequency bands given the limited selectivity of the antenna itself. If the receiver will be operated in an environment that will not have high-power adjacent frequency band transmitters in close proximity, then this filter may not be necessary and the position of the first amplifier and filter can be switched. Or the impact to the noise figure can be minimized by selecting a filter with a low insertion loss.

The second filter parameter is the *bandwidth*. Again, since no filter is ideal, typically a 3 dB bandwidth is specified. This indicates at which frequency(s) the attenuation will be 3 dB (or 50% of the signal power). However, these two parameters cannot completely describe most filters but only provide some insight into their performance, as shown in Figure 4.3.

A goal in filter design is to provide sharp transitions between the desired (passband) and undesired (cutoff) frequencies while maintaining a minimal insertion loss. Depending on the practical implementation of the filters [options include cavity, surface acoustic wave (SAW), ceramic, or lumped elements (resistors, capacitors, and filters)], this can be done by increasing the number of sections/elements within the design.

4.2.3 Amplifier

Amplification is the process that increases the signal magnitude. Thus, an amplifier is a component that does just that. Unlike most filters, an amplifier is an active component and requires power to accomplish its function. Note that the ideal amplifier would only increase the amplitude of the signal. However, any commercial amplifier will not only increase the amplitude but also add noise to the resulting signal. The goal, of course, is to have a component that amplifies the signal and adds minimal noise.

The fundamental parameters used to describe an amplifier are

1. *gain*, usually expressed in dB, and often assumed constant over a

2. *specified frequency range*; and a

3. *noise figure*, again usually expressed in dB, and indicative of the amount of noise that will be added to the signal being amplified.

Note that this discussion simplifies the practical amplifier. We are assuming the amplifier is a packaged device, ignoring the actual fabrication. Further, parameters such as the third-order intercept point, power requirements, and maximum power handling are all additional factors that could be considered but are neglected to simplify the discussion.

Also the amplification shown in Figure 4.2 shows a single amplifier capable of 50 dB gain. It would be unusual to have a single amplifier capable of such gain. What is represented as a single amplifier in Figure 4.2 can be constructed of cascaded multistage amplifiers.

The goal of the amplifier is to raise the extremely weak incident signal to a level practical for analog-to-digital conversion. Thus, the amount of amplification is based on the specific ADC and will be discussed in that section. Further, there is typically a distribution of amplification or gain across different frequencies for reasons that will become obvious in the next subsection.

4.2.4 Mixer/Local Oscillator

The basic function of the mixer/local oscillator combination is to translate the input 1575.42 MHz RF carrier to a lower intermediate frequency (IF) and preserve the modulated signal structure. The most obvious reason for this is to bring the frequency to usable ranges in which to operate on the signal, in particular perform the analog-to-digital conversion. However, there are fewer obvious reasons for the frequency translation, as is discussed within this subsection.

The design illustrated in Figure 4.2 utilizes a single stage of analog frequency translation. However, it is possible to utilize multiple stages of analog frequency translation in a single front-end design. The choice is a design trade-off based on the components available and their individual specifications. The focus of this section is to illustrate the functionality of the single-stage approach shown in Figure 4.2.

First, it is important to discuss the individual components. The local oscillator for GNSS front-end designs is typically a combination of components. Most crystal oscillators, either standalone or temperature compensated/ovenized for greater stability, are not capable of generating the desired local oscillator frequency for the L1 GNSS signal. Thus, a phase lock loop (PLL) is combined with the crystal to achieve the desired higher frequency of the local oscillator. In addition, it is common practice that the local oscillator be divided down to serve as the sampling clock, as shown in Figure 4.2. This is an important aspect as a single frequency source, and any associated frequency error/drift, will serve as the basis for the receiver.

The mixer operates through the trigonometric identity expressed as

$$\cos(\omega_1 t)\cos(\omega_2 t) = \tfrac{1}{2}\cos\big((\omega_1 - \omega_2)t\big) + \tfrac{1}{2}\cos\big((\omega_1 + \omega_2)t\big). \qquad (4.3)$$

It is possible to use the front-end design in Figure 4.2 as an example of the mixing process. In this case ω_1 equals the GNSS L1 center frequency 1575.42 MHz and the desired IF is 47.74 MHz, then the desired local oscillator frequency ω_2 would be $(1575.42 - 47.74)$ MHz $= 1527.68$ MHz. Any modulation, such as the GNSS spreading codes and navigation data, can be simply expressed as a time-varying multiplier:

$$d(t)\cos(\omega_1 t)\cos(\omega_2 t) = \tfrac{d(t)}{2}\cos\big((\omega_1 - \omega_2)t\big) + \tfrac{d(t)}{2}\cos\big((\omega_1 + \omega_2)t\big). \quad (4.4)$$

In this case it is obvious that the output of the mixer will be the sum and difference frequencies. Of interest here is the difference frequency, which is at the desired IF. The sum frequency is simply a consequence in this case, and the second filter depicted in the cascade of Figure 4.2, which follows the mixer, is used to select only the desired difference frequency.

Note that in Figure 4.2 a bandpass filter is used for this process. However, given the fact that the goal is to simply remove the sum component, a lowpass filter should be more than sufficient. In many cases this is true; however, Equations (4.3) and (4.4) present a simplified model of a mixer, which in reality is more complicated. Mixer parameters include conversion loss, isolation, dynamic range, and intermodulation. In this case the bandpass filter is selected to minimize any complications from intermodulation products that result from the mixing. For this simplified discussion, only the straightforward model of the mixer is presented.

With the combination local oscillator/mixer it is now possible to translate the RF carrier to a lower IF. It has been eluded to above that this is required for the analog-to-digital conversion process, but is that the only reason? Are there other reasons as to why frequency translation is important in GNSS receivers? The answer is "yes" with two immediate additional justifications for the frequency translation.

The first is the quality and cost of the component. The goal of this text is to develop software GNSS receivers for the narrowband L1 signals, with the definition of narrowband being 2–8 MHz (see Problem 7). It is important to recognize that it can be quite difficult to fabricate narrowband filters at high frequency.

Denote the quality factor of the filter by Q, the center frequency of the filter (1575.42 MHz for GNSS L1) by f_{center}, and the bandwidth of that filter by BW. Then the quality of a filter is defined by

$$Q = \frac{f_{center}}{BW}. \qquad (4.5)$$

If we assume a 3 dB bandwidth and desire a filter to capture the main lobe of the GPS spectrum (2.046 MHz wide), the Q factor for such a filter comes out to be 770—an extremely high value. To put things in perspective, fabrication of most commercial filters (although it does depend on the technology) sets a minimal bandwidth of 2% of the center frequency. This corresponds to a Q value of 50, significantly less than the 770 computed above.

However, perform the same computation at the resulting 47.74 MHz IF from the design in Figure 4.2. That Q factor is 47.74/2.046 or 23.33, which is a much more realizable filter. Thus, the frequency translation to IF allows higher frequency selectivity with less costly/complex components.

The second additional factor motivating frequency translation is feedback. The amount of amplification in the RF chain is tremendous; over 100 dB of gain is applied. If this is all attempted at a single frequency, then it is highly likely that feedback will become an issue unless meticulous shielding and spatial separation across the RF chain is implemented. Otherwise, if the 100+ dB of gain were applied all within the 1575.42 MHz band even with quality RF cabling between components, it is unlikely to prevent feedback within the amplification stages in the RF chain. Utilizing multiple stages within a front-end design allows the gain to be distributed across frequency. For example, in the single-stage downconversion depicted in Figure 4.2 the gain within the RF chain is split between the RF and IF paths. Thus the level of shielding and potential for feedback are reduced as the output of the lower-frequency amplifiers cannot feedback into the input of the higher-frequency amplifiers.

4.2.5 Analog-to-Digital Converter

The final component in the front-end path is the analog-to-digital converter. This device is responsible for the conversion of the analog signal to digital samples. There is a wide variety of ADCs available on the market, with a dizzying set of parameters for each. Consider, for example, the Texas Instruments ADS830 ADC, see focus.ti.com/lit/ds/symlink/ads830.pdf. Such an ADC has an overwhelming number of parameters, the majority of which are not discussed here. An application note can help users sort out the various parameters associated with ADCs; see Anonymous (2000).

The key parameters to consider for this discussion are the *number of bits*, the *maximum sampling frequency*, the *analog input bandwidth*, and the *analog input range*.

The CDMA nature of the GNSS signal requires very little dynamic range from the sampled signal. It has been shown that if single bit sampling is used, then degradation in the resulting processing is less than 2 dB; see Bastide et al. (2003). Further, if conservative 2- or more bit sampling is utilized with proper quantization, the degradation is less than 1 dB. The minimum number of bits on most commercial ADCs is 8 as is the case for the ADS830. Thus, in designing a GNSS front end, it is most convenient to either utilize a hard limiter to obtain a single bit or use a commercial ADC taking all or just a subset of the resulting bits of each sample. It is also important to recognize that if multibit sampling is employed, then some form of gain control must be implemented to provide proper quantization.

One might ask if the penalty for using single bit sampling is less than 2 dB, why any front end would utilize multibit sampling and then incorporate the overhead associated with automatic gain control? The key point to remember is that the less-than 2 dB penalty is for the ideal case. If, for example, there exists narrowband interference within the GNSS L1, then single bit sampling will be captured by the interference source and prevent GNSS processing. Thus, although the theoretical penalty for single-bit sampling is less than 2 dB, the nature of the operating environment may dictate the need for multibit sampling.

The maximum sampling frequency is an interesting parameter. This frequency needs to accommodate the bandwidth of the desired signal. Continuing to use the ADS830 part as an example, the maximum sampling frequency is 60 MHz and thus can provide a resulting sampling bandwidth of 30 MHz, more than sufficient for the narrowband L1 navigation signals. However, recognize that the IF in Figure 4.2 is at 47.74 MHz, which is greater than the resulting [0–30] MHz sampled information bandwidth. In this case, the sampling process acts as a second frequency translation stage.

Although the ADC has a sampling frequency that provides an upper limit of 30 MHz on the resulting sampling bandwidth, then analog input bandwidth of this ADC really determines what signals will be captured. For the ADS830, this value is an impressive 300 MHz. What this means is that any frequency component input to the ADC up to 300 MHz will be aliased according the sampling theorem.

Should the analog input bandwidth have been as high as 1.6 GHz, which is not impossible (see pdfserv.maxim-ic.com/en/ds/MAX104.pdf), then it would be possible to directly sample and alias the original RF signal. Such an implementation has been demonstrated [Akos (1997) and Akos et al. (1999)] yet there remain many technical hurdles to overcome with such an approach. The approach outlined does provide the means to compute an appropriate sampling frequency and the resulting sampling IF.

Based on the preceding discussion, the role of the final filter in the RF chain becomes clear. It must be a bandpass filter and limit the band to only those frequencies to be preserved through the sampling process. Recognize that the aliasing does not only occur for the desired IF, but all frequencies within the analog input bandwidth of the ADC. Thus, it is critical for minimal noise that the last filter prior to the ADC allows only those frequencies of interest and attenuates all others within the analog input bandwidth.

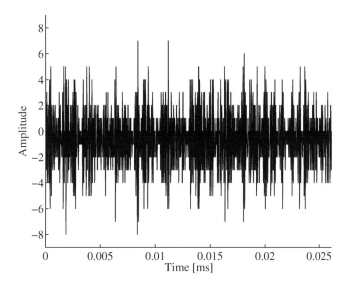

FIGURE 4.4. Time domain. Representation of 1000 samples.

Referring to the design in Figure 4.2, a sampling frequency of 38.192 MHz is used for the 47.74 MHz IF, and this provides a final digital frequency translation to an IF of 9.548 MHz. The resulting sampled information bandwidth of [0–38.192/2] MHz provides more.

The final ADC parameter to be discussed is the analog input range. This range defines the voltage range for which the quantization will be distributed across. In the case of the ADS830 part, the minimum analog input range is 1 V peak-to-peak. Assuming a 50 Ω load, which is traditional in radio frequency design, a 1 V signal corresponds to −17 dBW. Thus, it should now be clear why the amplification within the RF chain is needed. It had been mentioned that it was to provide suitable signal levels. The combination of thermal noise and received signal will be simply too weak to exercise the bits in this or any ADC. So a goal of the amplifiers is to increase the received signal level, which again is dominated by the thermal noise, to exercise the full range of the ADC.

Although not depicted in Figure 4.2, the final amplifier in many GNSS front-end designs will be a variable gain with a feedback signal resulting from processing implemented after the ADC. This is implemented and known within most GNSS receivers as *automatic gain control* (AGC). The goal of the front end is to exercise all available bits with the ADC. Thus, if the gain is insufficient to do so and this is determined by monitoring the sampled data stream, the gain can be increased. Alternatively, if the gain is too high such that the outer ADC bins have an overwhelming number of samples, then the gain can be decreased. Lastly, as will be discussed when the sampled data are presented, the AGC can be steered by the expected distribution of the sample bins. In this way, the front end can attempt to minimize the impact of narrowband interference.

In summary, the object of the bulk of the components within the front end is to condition the voltage incident on the antenna for sampling by the ADC. In order to

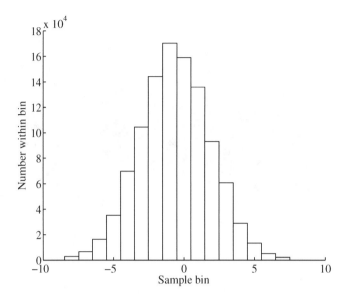

FIGURE 4.5. Histogram of 1,048,576 data samples.

accomplish this for most ADCs there are three basis functions which must be accomplished. These are *amplification*, *frequency translation/downconversion*, and *filtering*. These prepare the signal for analog-to-digital conversion, which results in the samples to be processed within the software receiver.

4.3   Resulting Sampled Data

Now that the operation/functionality of the GNSS front end has been described, it is worthwhile to highlight the resulting data that have been collected from the front-end design depicted in Figure 4.2 and have been included on the media with the book.

Again, the important parameters for the signal processing are

– Sampling frequency: 38.192 MHz
– Intermediate frequency: 9.548 MHz, and
– Four-bit samples.

The above parameters provide all the necessary information for the operation of the signal processing algorithms. Some other items, such as the time and date and approximate location of the data collection, can speed the acquisition as will be discussed, but are not required.

What can be done is to show the resulting digital samples in typical representations. Thus, in Figures 4.4, 4.5, and 4.6, a time domain, histogram, and frequency domain depiction of the collected data are illustrated, respectively.

In the time domain depiction, no discernable structure is visible despite the 9.548 MHz IF for the collected GPS data. In the histogram, it is obvious that all four bits of the ADC are being triggered based on the 16 levels present within

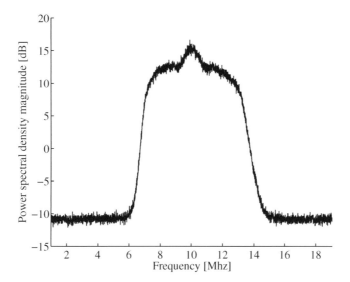

FIGURE 4.6. Frequency domain. Representation of 1,048,576 samples of GPS L1 data.

the histogram. Also the histogram bears a strong resemblance to the probability density function for a Gaussian random variable, which would be expected for white thermal noise. These plots follow what would be expected based on the frequency domain depiction in Figure 4.1 where the thermal noise would dominate the resulting samples.

However, the frequency domain depiction does not resemble Figure 4.1; rather some obvious structure is present. This structure is best explained by building on Figure 4.1, as shown in Figure 4.7.

In this figure, obvious changes have been made to better correspond to the frequency domain depiction of the collected data file.

First, the "noise level" is not white, or flat and uniform as a function of frequency, but has some definite structure. This is a result of the final bandpass filter prior to sampling. This 6 MHz-wide bandpass filter shapes the spectrum of the analog signal to be sampled. Thus, the filter shape has been added within Figure 4.7.

Second, there is an obvious "bulge" within the center of the passband right at the resulting IF of the GPS translated signal. This actually appears to be the main 2.046 MHz lobe of the sinc spectrum of the signal itself. With the specified received signal power level so much lower than the expected thermal noise, how can there be any discernable structure from the satellite signal from a data set collected with a traditional hemispherical antenna? The explanation has two components:

– the individual received satellite signal power is currently higher than the minimum specified (as shown in Figure 4.7); and

– the CDMA nature of the GPS system has all the satellite signal power overlaid at the resulting IF; thus, the spectrum shows the summation of all the visible satellite signal power.

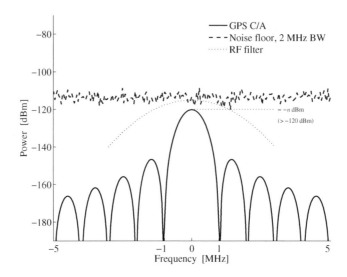

FIGURE 4.7. Improved frequency domain depiction. Center frequency 1575.42 MHz.

These two factors bring about some, although minimal, structure from the satellite signal spectrum into the frequency domain representation of the collected data.

For the most part, the data indeed resemble thermal noise, and all the traditional GNSS signal processing is required to acquire, track, and utilize the navigation transmission.

4.4   GNSS Front-End ASIC

The final section of this chapter presents the current state of the art for GNSS front-end designs. The implementation in Figure 4.2 has been built from expensive discrete components. Although this provides a high-quality front-end suitable for a lab instrument or environment, the cost for such an implementation can approach $5000. With the cost of handheld GPS receivers now well below $100, an alternative to such a design must exist.

The solution comes in the form of an integrated circuit. The bulk of the functionality of Figure 4.2 has been incorporated by multiple vendors into an ASIC that is typically smaller than 5×5 mm packages and utilizes less than 50 mW; see SiGe SE4110L, Nemerix NJ1006, and Texas Instruments TRF5101 data sheets. Such components strive to be as completely self-contained as possible, requiring only a minimal number of external components.

For example, consider the block diagram for the SiGe SE4110L component shown in Figure 4.8. This is an excellent example of a GPS front-end ASIC component. The complete data sheet for the SE4110L is included on the bundled DVD.

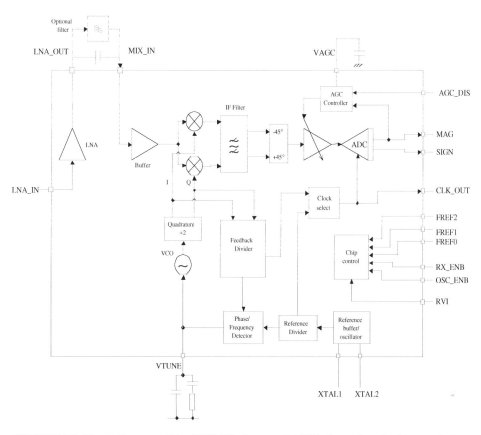

FIGURE 4.8. Block diagram of the SE4110L front-end ASIC. Reproduced with permission of SiGe Semiconductor, Inc.

Based on the discussion within the chapter, the underlying design of the component should be somewhat familiar. A single-frequency translation stage is utilized along automatic gain control functionality to support multiple bit sampling.

It is quite impressive to see the level of integration within such a component. It utilizes a traditional antenna input (although a passive antenna can be used with an internal LNA with noise figure of less than 2.5 dB) and provides 2-bit digital samples supporting a number of different clock frequencies for a variety of applications. This particular part is only 4 × 4 mm and draws less than 10 mA from a nominal 2.7–3.3 V supply. Such integration is even more impressive when one considers the gain required for processing the received signal power of the GPS and potential feedback issues.

Such development even further facilitates the wide scale deployment of satellite navigation technology at a relatively low cost.

The ASIC-based front end is just one of multiple options for converting the signal in space to a digital format suitable for the software based signal processing. The goal of this chapter is to provide some insight into the source of that data.

5

GNSS Receiver Operation Overview

5.1 Receiver Channels

The signal processing for satellite navigation systems is based on a channelized structure. This is true for both GPS and Galileo. This chapter provides an overview of the concept of a receiver channel and the processing that occurs. In later chapters the specifics of the signal and data processing are outlined.

Figure 5.1 gives an overview of a channel. Before allocating a channel to a satellite, the receiver must know which satellites are currently visible. There are two common ways of finding the initially visible satellites. One is referred to as *warm start* and the other is referred to as *cold start*.

Warm start In a warm start, the receiver combines information in the stored almanac data and the last position computed by the receiver. The almanac data is used to compute coarse positions of all satellites at the actual time. These positions are then combined with the receiver position in an algorithm computing which satellites should be visible. The warm start has at least two downsides. If the receiver has been moved far away since it was turned off (e.g., to another continent), the receiver position cannot be trusted and the found satellites do not match the actual visible satellites. Another case is that the almanac data can be outdated, so they cannot provide good satellite positions. In either case, the receiver has to make a cold start.

Cold start In a cold start, the receiver does not rely on any stored information. Instead it starts from scratch searching for satellites. The method of searching is referred to as acquisition and it is described in the following section.

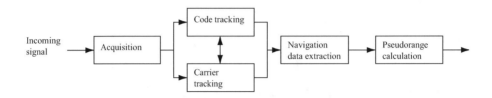

FIGURE 5.1. One receiver channel. The acquisition gives rough estimates of signal parameters. These parameters are refined by the two tracking blocks. After tracking, the navigation data can be extracted and pseudoranges can be computed.

An acquisition search through all possible satellites is quite time-consuming. That is, in fact, the reason why a warm start is preferred if possible.

5.1.1   Acquisition

The purpose of acquisition is to identify all satellites visible to the user. If a satellite is visible, the acquisition must determine the following two properties of the signal:

Frequency The frequency of the signal from a specific satellite can differ from its nominal value. In case of downconversion, the nominal frequency of the GPS signal on L1 corresponds to the IF. However, the signals are affected by the relative motion of the satellite, causing a Doppler effect. The Doppler frequency shift can—in the case of maximum velocity of the satellite combined with a very high user velocity—approach values as high as 10 kHz; see Tsui (2000), page 39. For a stationary receiver on Earth, the Doppler frequency shift will never exceed 5 kHz.

Code Phase The code phase denotes the point in the current data block where the C/A code begins. If a data block of 1 ms is examined, the data include an entire C/A code and thus one beginning of a C/A code.

Many different methods have been used: they are all in one way or another based on the GPS signal properties. The C/A code correlation properties are especially important; see Section 2.3.4.

The received signal s is a combination of signals from all n visible satellites

$$s(t) = s^1(t) + s^2(t) + \cdots + s^n(t). \tag{5.1}$$

When acquiring satellite k, the incoming signal s is multiplied with the local generated C/A code corresponding to the satellite k. The cross correlation between C/A codes for different satellites implies that signals from other satellites are nearly removed by this procedure. To avoid removing the desired signal component, the locally generated C/A code must be properly aligned in time, that is, have the correct code phase.

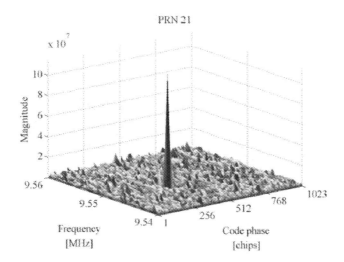

FIGURE 5.2. Acquisition plot for PRN 21. Signals originating from PRN 21 are present in the received signal. This is seen from the significant peak in the acquisition plot. The peak location is related to a C/A code phase and a frequency of the signal.

After multiplication with the locally generated code, the signal must be mixed with a locally generated carrier wave. This is done to remove the carrier wave of the received signal. To remove the carrier wave from the signal, the frequency of the locally generated signal must be close to the signal carrier frequency. As mentioned earlier, the frequency can change up to $\pm10\,\mathrm{kHz}$ from the nominal frequency, so different frequencies within this area must be tested. To identify whether or not a satellite is visible, it is sufficient to search the frequency in steps of 500 Hz resulting in 41 different frequencies in case of a fast-moving receiver and 21 in case of a static receiver; see Akos (1997), page 85.

After mixing with the locally generated carrier wave, all signal components are squared and summed providing a numerical value.

The acquisition procedure works as a search procedure. For each of the different frequencies 1023 different code phases are tried. When all possibilities for code phase and frequency are tried, a search for the maximum value is performed. If the maximum value exceeds a determined threshold, the satellite is acquired with the corresponding frequency and phase shift. Figure 5.2 shows a typical acquisition plot performed for a visible satellite. The plot shows a significant peak, which indicates high correlation.

Figure 5.3 shows a typical acquisition plot, performed for a satellite that is not currently visible at the GPS receiver. In this plot, all values are nearly identical, indicating low correlation.

5.1.2 Tracking

The main purpose of tracking is to refine the coarse values of code phase and frequency and to keep track of these as the signal properties change over time.

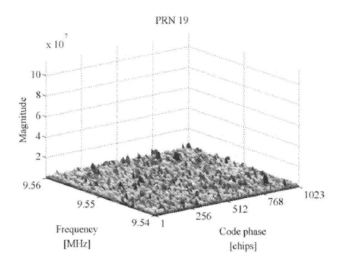

FIGURE 5.3. Acquisition plot for PRN 19. Signals originating from PRN 19 are clearly not present in the received signal as there is no sign of a peak in the acquisition plot.

The accuracy of the final value of the code phase is connected to the accuracy of the pseudorange computed later on. The tracking contains two parts, code tracking and carrier frequency/phase tracking:

Code tracking The code tracking is most often implemented as a delay lock loop (DLL) where three local codes (replicas) are generated and correlated with the incoming signal. These three replicas are referred to as the early, prompt, and late replica, respectively. The three codes are often separated by a half-chip length.

Carrier frequency/phase tracking The other part of the tracking is the carrier wave tracking. This tracking can be done in two ways: either by tracking the phase of the signal or by tracking the frequency.

The tracking is running continuously to follow the changes in frequency as a function of time. If the receiver loses track of a satellite, a new acquisition must be performed for that particular satellite.

5.1.3 Navigation Data Extraction

When the signals are properly tracked, the C/A code and the carrier wave can be removed from the signal, only leaving the navigation data bits. The value of a data bit is found by integrating over a navigation bit period of 20 ms. After reading about 30 s of data, the beginning of a subframe must be found in order to find the time when the data was transmitted from the satellite.

When the time of transmission is found, the ephemeris data for the satellite must be decoded. This is used later on to compute the position of the satellite at the time of transmission.

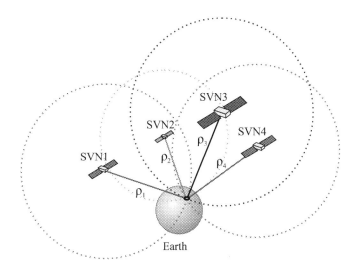

FIGURE 5.4. The basic principle of GNSS positioning. With known position of four satellites SVNi and signal travel distance ρ_i, the user position can be computed.

The last thing to do before making position computations is to compute pseudoranges. The pseudoranges are computed based on the time of transmission from the satellite and the time of arrival at the receiver. The time of arrival is based on the beginning of the subframe.

5.2   Computation of Position

The final task of the receiver is to compute a user position. The position is computed from pseudoranges and satellite positions found from ephemeris data. Figure 5.4 gives an impression of the method of position computation using GPS.

Section 8.5 gives a detailed description of the computational algorithm.

6
Acquisition

The present and the following chapters are based on signals recorded according to parameters described in Section 4.3. The theory can be applied similarly to records with a different selection of parameters.

6.1 Motivation

The purpose of acquisition is to determine visible satellites and coarse values of carrier frequency and code phase of the satellite signals.

The satellites are differentiated by the 32 different PRN sequences. The second parameter, code phase, is the time alignment of the PRN code in the current block of data. It is necessary to know the code phase in order to generate a local PRN code that is perfectly aligned with the incoming code. Only when this is the case, the incoming code can be removed from the signal. PRN codes have high correlation only for zero lag. That is, the two signals must be perfectly aligned to remove the incoming code. The third parameter is the carrier frequency, which in case of downconversion corresponds to the IF. The IF should be known from the L1 carrier frequency of 1575.42 MHz and from the mixers in the downconverter. However, the frequency can deviate from the expected value. The line-of-sight velocity of the satellite (with respect to the receiver) causes a *Doppler effect* resulting in a higher or lower frequency. In the worst case, the frequency can deviate up to ± 10 kHz. It is important to know the frequency of the signal to be able to generate a local carrier signal. This signal is used to remove the incoming carrier from the signal. In most cases it is sufficient to search the frequencies such that the maximum error will be less than or equal to 500 Hz; see Akos (1997).

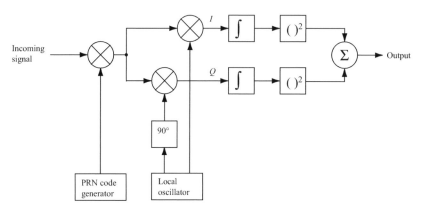

FIGURE 6.1. Block diagram of the serial search algorithm.

In an ordinary receiver, the acquisition is usually performed in an application-specific integrated circuit (ASIC). In a software receiver, it is implemented in software. The following sections describe the theory behind three standard methods of acquisition to demonstrate the possibility of implementing an efficient method in a software receiver.

6.2   Serial Search Acquisition

Serial search acquisition is an often-used method for acquisition in code-division multiple access systems (CDMA). GPS is a CDMA system. Figure 6.1 is a block diagram of the serial search algorithm.

As seen in Figure 6.1, the algorithm is based on multiplication of locally generated PRN code sequences and locally generated carrier signals. The PRN generator generates a PRN sequence corresponding to a specific satellite. The generated sequence has a certain code phase, from 0 to 1022 chips. The incoming signal is initially multiplied by this locally generated PRN sequence. After multiplication with the PRN sequence, the signal is multiplied by a locally generated carrier signal. Multiplication with the locally generated carrier signal generates the in-phase signal I, and multiplication with a 90° phase-shifted version of the locally generated carrier signal generates the quadrature signal Q.

The I and Q signals are integrated over 1 ms, corresponding to the length of one C/A code, and finally squared and added. Ideally, the signal power should be located in the I part of the signal, as the C/A code is only modulated onto that. However, in this case the I signal generated at the satellite does not necessarily correspond to the demodulated I. This is because the phase of the received signal is unknown. So to be certain that the signal is detected, it is necessary to investigate both the I and the Q signal. The output is a value of correlation between the incoming signal and the locally generated signal. If a predefined threshold is exceeded, the frequency and code phase parameters are correct, and the parameters can be passed on to the tracking algorithms.

The serial search algorithm performs two different sweeps: a frequency sweep over all possible carrier frequencies of IF $\pm 10\,$kHz in steps of 500 Hz and a code phase sweep over all 1023 different code phases. All in all, this sums up to a total of

$$\underbrace{1023}_{\text{code phases}} \underbrace{\left(2\frac{10,000}{500} + 1 \right)}_{\text{frequencies}} = 1023 \cdot 41 = 41{,}943 \text{ combinations.} \qquad (6.1)$$

Obviously, this is a very large number of combinations. This exhausting search routine also tends to be the main weakness of the serial search acquisition.

The implementation of the serial search acquisition method is very straightforward. The algorithm can be implemented directly based on the block diagram of the method as shown in Figure 6.1.

6.2.1 PRN Sequence Generation

Figure 6.1 shows that the first task in the serial search acquisition method is to multiply the incoming signal with the locally generated PRN sequence. This of course involves the generation of this PRN sequence. Instead of generating PRN sequences every time the algorithm is executed, all possible PRN sequences are generated offline. The 32 different PRN sequences are generated by the PRN generator implemented according to Figure 2.5.

The PRN code generator is implemented using the binary values 0 and 1. However, in the signal processing algorithms it is more convenient to represent the codes with a polar non-return-to-zero representation.

With 32 generated PRN sequences, all possible sequences originating from GPS satellites are created. However, as mentioned in the theory of serial search acquisition, the method involves multiplication with all possible shifted versions of the PRN codes. That is, besides saving the 32 possible PRN codes all possible shifted versions should also be saved. This sums up to a total of 32,736 different PRN codes. To make the multiplication between one of the generated PRN codes and the incoming signal possible, the code has to be *sampled* with 38.192 MHz, like the received signal. This sampling converts the length of a PRN sequence from 1023 to 38,192.

6.2.2 Carrier Generation

The second step is multiplication with a locally generated carrier wave. The carrier generator must generate two carrier signals with a phase difference of 90°, corresponding to a cosine and a sine wave. The carrier must have a frequency corresponding to the IF $\pm$ the frequency step according to the examined frequency area. It must be sampled with the sampling frequency of 38.192 MHz and have a length of 1 ms. A complex signal is generated using the natural exponential function $e^{j2\pi f}$.

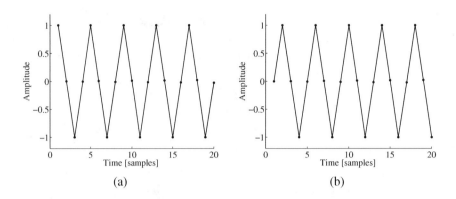

FIGURE 6.2. Locally generated cosine (a) and sine (b) waves.

In MATLAB the 38,192 sample-long sequence, corresponding to 1 ms, is generated as carrier = exp(j * 2 * pi * fc * ts * nn); where fc is the IF plus current frequency offset, ts is the sampling period $t_s = 1/f_s$, and nn is an array with numbers from 0 to 38,191. Now the sine and cosine waves are generated as cosine = real(carrier); and sine = imag(carrier);.

Figure 6.2 shows the first 20 samples of the resulting cosine and sine waves when using a sampling frequency $f_s = 38.192\,\text{MHz}$ and letting $f_c = 9.548\,\text{MHz}$.

6.2.3   Integration and Squaring

The last parts of the serial search algorithm involve an integration and a squaring of the two results of the multiplications with the cosine and sine signals, respectively. The squaring is introduced to obtain the signal power. The integration is simply a summation of all 38,192 points corresponding to the length of the processed data. The squaring is then performed on the result of the summation. The final step is to add the two values from the I arm and the Q arm. If the locally generated code is well aligned with the code in the incoming signal, and the frequency of the locally generated carrier matches the frequency of the incoming signal, the output will be significantly higher than if any of these criteria were not fulfilled. Figure 6.3 shows the output from the serial search acquisition method. Figure 6.3a is the output when performing the acquisition on a satellite that is not visible and Figure 6.3b when performing the acquisition on a satellite that is visible. PRN 19 is arbitrarily chosen as a nonvisible satellite.

6.3   Parallel Frequency Space Search Acquisition

The serial search acquisition method showed that it is a very time-consuming procedure to search sequentially through all possible values of the two parameters frequency and code phase. If any of the two parameters could be eliminated from the search procedure or if possible implemented in parallel, the performance of the procedure would increase significantly.

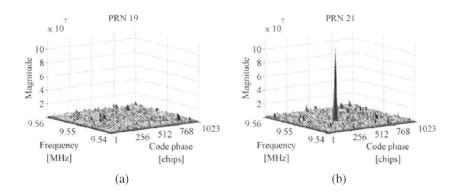

FIGURE 6.3. Output from serial search acquisition. (a) PRN 19 is not visible so no peak is present. (b) PRN 21 is visible so a significant peak is present. The peak occurs at C/A code phase = 359 chips and frequency = 9.5475 MHz.

As the name *parallel frequency space search acquisition* implies, this second method of acquisition parallelizes the search for the one parameter. This method utilizes the Fourier transform to perform a transformation from the time domain into the frequency domain. Figure 6.4 is a block diagram of the parallel frequency space search algorithm.

The incoming signal is multiplied by a locally generated PRN sequence, with a code corresponding to a specific satellite and a code phase between 0 and 1022 chips. The resulting signal is transformed into the frequency domain by a Fourier transform. The Fourier transform could be implemented as a discrete Fourier transform (DFT) or a fast Fourier transform (FFT). The FFT is the faster of the two; but it requires an input sequence with a radix-2 length, that is, 2^n, where n takes positive integer value.

Figure 6.5 illustrates the result of multiplying the incoming signal with a perfectly aligned locally generated PRN sequence. The result is a continuous wave signal. Of course, this only happens when the locally generated PRN code is perfectly aligned with the code in the incoming signal. If the incoming signal contains signal components from other satellites, these components will be minimized as a result of the cross-correlation properties of the PRN sequences.

In parallel frequency space search acquisition, the upper signal in Figure 6.5 is the input to the Fourier transform function. With a perfectly aligned PRN code, the output of the Fourier transform will show a distinct peak in magnitude. The

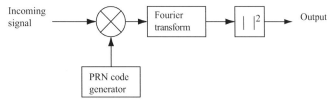

FIGURE 6.4. Block diagram of the parallel frequency space search algorithm.

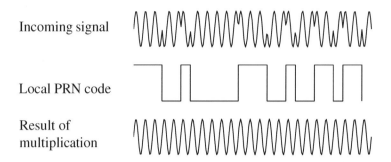

Incoming signal

Local PRN code

Result of
multiplication

FIGURE 6.5. PRN code demodulation. At the top the PRN code is modulated onto the carrier wave. In the middle is a perfectly aligned PRN sequence. At the bottom we recover the continuous wave after multiplying the incoming signal with the perfectly aligned PRN sequence.

peak will be located at the frequency index corresponding to the frequency of the continuous-wave signal and thereby the frequency of the carrier wave signal.

The accuracy of the determined frequency depends on the length of the DFT. The length corresponds to the number of samples in the analyzed data. If 1 ms of data is analyzed, the number of samples can be found as 1/1000 of the sampling frequency. That is, if the sampling frequency is $f_s = 10$ MHz, the number of samples is $N = 10,000$.

With a DFT length of 10,000, the first $N/2$ output samples represent the frequencies from 0 to $\frac{f_s}{2}$ Hz. That is, the frequency resolution of the output is

$$\Delta f = \frac{f_s/2}{N/2} = \frac{f_s}{N}. \tag{6.2}$$

With a sampling frequency of $f_s = 10$ MHz the resulting frequency resolution is

$$\Delta f = \frac{10\,\text{MHz}}{10,000} = 1\,\text{kHz}. \tag{6.3}$$

In this case, the accuracy of the estimated carrier frequency is 1 kHz compared to the accuracy of 500 Hz in serial search acquisition.

Figure 6.6 shows the output as two power spectral density (PSD) plots. The PSD algorithm is using FFT for the acquisition. Figure 6.6a is the output of the Fourier transform with a perfectly aligned code phase. This can be noticed from the peak in the plot. Figure 6.6b shows the PSD plot for a not aligned code phase. Note the absence of a peak in the plot.

Where the serial search acquisition method steps through possible code phases and carrier frequencies, the parallel frequency space search acquisition only steps through the 1023 different code phases. This comes with the cost of a frequency domain transformation with each code phase. Depending on the implementation of the frequency domain transformation, it should be possible to make a faster implementation of this method compared to the serial search acquisition method.

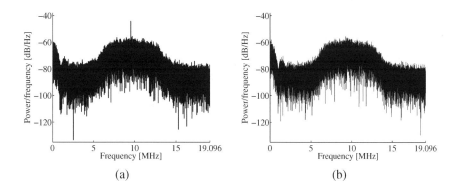

FIGURE 6.6. PSD plot of the incoming signal multiplied by a locally generated PRN code sequence. (a) When multiplying with a perfectly aligned PRN code, the output will show a peak at the carrier frequency. (b) When multiplying with a nonaligned code, the output will not show any peaks. The IF is 9.548 MHz. The Doppler frequency is the same on IF and RF, and thus also the difference between the IF and the peak frequency.

Like the serial search acquisition method, the implementation of the parallel frequency space search method is straightforward. The algorithm can be implemented directly based on the block diagram of the method shown in Figure 6.4.

The first part of this method is identical to the first part of the serial search method. That is, a locally generated PRN code must be multiplied with the incoming signal. After the code multiplication, the signal is transformed into the frequency domain through Fourier transform. An efficient tool for that is the fast Fourier transform (FFT). See Oppenheim & Schäfer (1999) for details on the FFT.

After transforming the signal into the frequency domain by means of the FFT algorithm, it becomes a complex signal. If the locally generated code is well aligned with the code in the incoming signal, the output from the FFT will have a peak at the IF plus Doppler offset frequency. To find the possible peak frequency the absolute value of all components are calculated. Figure 6.7 shows the output from the parallel frequency space search method.

6.4   Parallel Code Phase Search Acquisition

As seen from Equation (6.1), the amount of search steps in the code phase dimension is significantly larger than that of the frequency dimension (1023 compared to 41). The previous method parallelized the frequency space search eliminating the necessity of searching through the 41 possible frequencies. If the acquisition could be parallelized in the code phase dimension, only 41 steps should be performed compared to the 1023 in the parallel frequency space search acquisition algorithm.

A recent method in GPS signal acquisition utilizes the before-mentioned advantage of parallelizing the code phase search. This method is simply referred to as parallel code phase search acquisition. In the following we describe the theory behind this method.

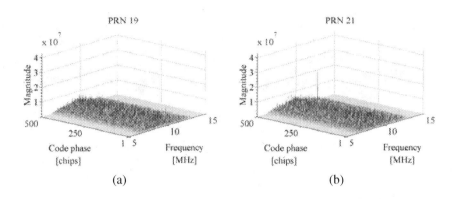

(a)   (b)

FIGURE 6.7. Output from parallel frequency space search acquisition. The figure only includes the first 500 chip shifts and the frequency band from 5–15 MHz. (a) PRN 19 is not visible so no significant peaks are present in the spectrum. (b) PRN 21 is visible so a significant peak is present in the spectrum. The peak is situated at code phase 359 chips and frequency 9.548 MHz.

The goal of the acquisition is to perform a correlation with the incoming signal and a PRN code. Instead of multiplying the input signal with a PRN code with 1023 different code phases as done in the serial search acquisition method, it is more convenient to make a circular cross correlation between the input and the PRN code without shifted code phase. In the following, a method of performing circular correlation through Fourier transforms will be described; see Oppenheim & Schäfer (1999), page 746, and Tsui (2000), page 140.

The discrete Fourier transforms of the finite length sequences $x(n)$ and $y(n)$ both with length N are computed as

$$X(k) = \sum_{n=0}^{N-1} x(n)e^{-j2\pi kn/N} \quad \text{and} \quad Y(k) = \sum_{n=0}^{N-1} y(n)e^{-j2\pi kn/N}. \quad (6.4)$$

The circular cross-correlation sequence between two finite length sequences $x(n)$ and $y(n)$ both with length N and with periodic repetition is computed as

$$z(n) = \frac{1}{N}\sum_{m=0}^{N-1} x(m)y(m+n) = \frac{1}{N}\sum_{m=0}^{N-1} x(-m)y(m-n). \quad (6.5)$$

In the following we will omit the scaling factor $\frac{1}{N}$.

The discrete N-point Fourier transform of $z(n)$ can be expressed as

$$Z(k) = \sum_{n=0}^{N-1}\sum_{m=0}^{N-1} x(-m)y(m-n)e^{-j2\pi kn/N}$$

$$= \sum_{m=0}^{N-1} x(m)e^{j2\pi km/N} \sum_{n=0}^{N-1} y(m+n)e^{-j2\pi k(m+n)/N} = X^*(k)Y(k), \quad (6.6)$$

where $X^*(k)$ is the complex conjugate of $X(k)$.

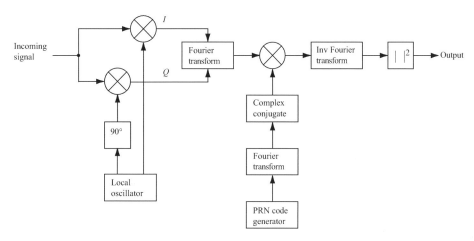

FIGURE 6.8. Block diagram of the parallel code phase search algorithm.

When the frequency domain representation of the cross correlation is found, the time-domain representation can be found through inverse Fourier transform.

Figure 6.8 is a block diagram of the parallel code phase search algorithm. The incoming signal is multiplied by a locally generated carrier signal. Multiplication with the signal generates the I signal, and multiplication with a 90° phase-shifted version of the signal generates the Q signal. The I and Q signals are combined to form a complex input signal $x(n) = I(n) + jQ(n)$ to the DFT function.

The generated PRN code is transformed into the frequency domain and the result is complex conjugated.

The Fourier transform of the input is multiplied with the Fourier transform of the PRN code. The result of the multiplication is transformed into the time domain by an inverse Fourier transform. The absolute value of the output of the inverse Fourier transform represents the correlation between the input and the PRN code. If a peak is present in the correlation, the index of this peak marks the PRN code phase of the incoming signal.

Compared to the previous acquisition methods, the parallel code phase search acquisition method has cut down the search space to the 41 different carrier frequencies. The Fourier transform of the generated PRN code must only be performed once for each acquisition. For each of the 41 frequencies we perform one Fourier transform and one inverse Fourier transform, so the computational efficiency of the method depends on the implementation of these functions. The accuracy of the parameters estimated by this acquisition method regards the frequency similar to the serial search method. The PRN code phase, however, is more accurate compared to the other methods as it gives a correlation value for each sampled code phase. That is, if the sampling frequency is 10 MHz, a sampled PRN code has 10,000 samples, so the accuracy of the code phase can have 10,000 different values instead of 1023.

In the same way as with the other acquisition methods, the implementation of this one is straightforward, as it can be implemented directly based on the block diagram shown in Figure 6.8.

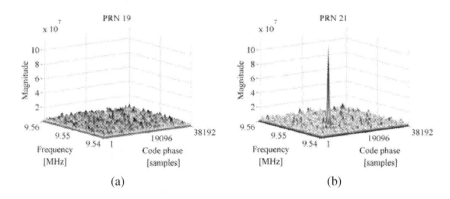

FIGURE 6.9. Output from parallel code phase search acquisition. (a) PRN 19 is not visible so no peak is present. (b) PRN 21 is visible so a significant peak is present. The peak is at code phase 13404 samples and frequency 9.5475 MHz.

As seen in Figure 6.8, this method involves almost no new blocks compared to the previous two. As a result of that, many elements can be reused in the implementation. One difference in this method is that only one PRN code should be generated for each acquisition. That is, it is not necessary to take the 1023 different code phases into account.

The first step is to multiply the incoming signal with a locally generated cosine and sine carrier wave, respectively, giving an I and a Q signal component. These two are combined as a complex input to the Fourier transform. The result of the Fourier transform is multiplied with the result of the lower branch of the block diagram in Figure 6.8. This signal is created as follows.

The PRN generator generates a code with no code phase. As in the implementation of the other acquisition algorithms, the code generation is performed *off-line*. The next step performs a Fourier transform of the PRN code, and the result is complex conjugated.

The result of the before-mentioned multiplication is given as input to an inverse Fourier transform implemented as the built-in function IFFT in MATLAB. This function has properties similar to the FFT function regarding execution time.

As mentioned in implementation of the parallel frequency space search acquisition method above, the output from an FFT is complex. This is also the case for the IFFT, so the output needs the same processing as in that case. That is, the absolute value is computed for all components. Figure 6.9 shows the output from the parallel code phase space search method.

6.5 Data Size

The selection of data size used for acquisition can be based on different criteria. The first issue to consider is the effect of navigation data bit transitions. None of the described algorithms ignores these transitions if they occur during the period of acquisition. So to guarantee optimal performance of the acquisition algorithms, it must be ensured that no data transitions occur in the analyzed data sequence.

TABLE 6.1. Execution time for each of the three implemented acquisition algorithms

Algorithm	Execution time	Repetitions	Complexity
Serial search	87	41,943	Low
Parallel frequency space search	10	1023	Medium
Parallel code phase search	1	41	High

As mentioned earlier, the navigation data are transmitted with a rate of 50 bps. This results in possible data bit transitions every 20 ms. If, for instance, 10 ms of data is used for the acquisition algorithm, it might include a bit transition. In fact, there is almost a 50% possibility that it does include a bit transition. (Not exactly 50% because two consecutive data bits might have identical values.) However, if acquisition is performed on two consecutive sequences, each with the length of 10 ms, at least one of the sequences will not include a data transition.

The second issue to consider when selecting the data size used for acquisition is the probability of making a successful acquisition. This issue can be discussed from the idea that the probability of detecting the correct parameters for a certain satellite increases with the amount of analyzed data.

The third issue is the computational demands as a function of the length of data to be analyzed. This is actually the counterpart to the previous issue, as the computations get slower when the sequence gets longer.

The choice of data size used for acquisition has to be a compromise based on the three issues just described. If the issue with data transitions have to be considered it might be necessary to run the acquisition algorithm twice for each acquisition. To ensure good probability of successful acquisition, the data length cannot be too short. However, it cannot be too long because this will cause the computations to be too heavy and time-consuming.

A compromise that will be used involves a data length of 1 ms for the acquisition algorithms. One ms corresponds exactly to the length of one complete C/A code, so it also simplifies the algorithm, making it unnecessary to duplicate the code. The data length can hardly be shorter, because this would involve correlation with an incomplete code. It could be longer, but as mentioned this would decrease the computational performance of the algorithm. To ensure that satellites will be acquired even though a data bit transition occurs in the analyzed data sequence, the algorithm can be run a second time if the first acquisition is unsuccessful.

6.6   Execution Time

As mentioned in the theory behind the acquisition methods, the theoretical performance regarding the computational demands are different between the three methods. So the execution time for each of the three implemented methods will

TABLE 6.2. Results of acquisition of PRN 21 with the three different acquisition algorithms; see captions of Figures 6.3, 6.7, and 6.9.

Search algorithm	Frequency [MHz]	Code phase
Serial	9.5475	359 chips
Parallel frequency space	9.548	359 chips
Parallel code phase	9.5475	13,404 samples

be measured to be used as a parameter for choosing the right algorithm in the receiver. The execution time is measured using the tic and toc functions in MAT-LAB. An average PC (Pentium 4, 2.8 GHz) was used for execution time measurements, all measurements are made 10 times, and the mean of these is computed. So the absolute value of the execution times is only approximate as none of these is optimized to run in realtime. The relative measures, however, should indicate which one will have the biggest potential for being implemented in realtime.

The results of the execution-time measurements can be seen in Table 6.1. The table also includes the number of repetitions or combinations the algorithm has to perform and the computational complexity of each of these repetitions.

Note that all relative measures are based on MATLAB performance. Implementation in other languages/environments may change the performance drastically.

6.7   Parameter Estimation

Another parameter that could be used for choosing between the three algorithms is the performance regarding precision of the result. The results from acquisition of satellite 25 using the three acquisition algorithms are shown in Table 6.2.

From Table 6.2 it is evident that all algorithms find the right frequency of the signal. The parallel frequency space search algorithm, however, estimates the frequency to be 9.548 MHz compared to 9.5475 MHz estimated by the other two.

7
Carrier and Code Tracking

7.1   Motivation

The acquisition provides only rough estimates of the frequency and code phase parameters. The main purpose of tracking is to refine these values, keep track, and demodulate the navigation data from the specific satellite (and provide an estimate of the pseudorange). A basic demodulation scheme is shown in Figure 7.1.

The figure shows the scheme used to demodulate the input signal to obtain the navigation message. First, the input signal is multiplied with a carrier replica. This is done to wipe off the carrier wave from the signal. In the next step, the signal is multiplied with a code replica, and the output of this multiplication gives the navigation message. So the tracking module has to generate two replicas, one for the carrier and one for the code, to perfectly track and demodulate the signal of one satellite. In the following, a detailed description of the demodulation scheme is conducted.

7.2   Demodulation

Let f_{L1} and f_{L2} be the carrier frequencies of L1 and L2 for the signal transmitted from satellite k with powers P_C, P_{PL1}, and P_{PL2} for C/A or P code. The C/A code sequence is $C^k(t)$ and the P(Y) code sequence is $P^k(t)$. If the navigation data sequence is called $D^k(t)$, the total signal is given as

$$s^k(t) = \sqrt{2P_C}C^k(t)D^k(t)\cos(2\pi f_{L1}t) + \sqrt{2P_{PL1}}P^k(t)D^k(t)\sin(2\pi f_{L1}t)$$
$$+ \sqrt{2P_{PL2}}P^k(t)D^k(t)\sin(2\pi f_{L2}t). \quad (7.1)$$

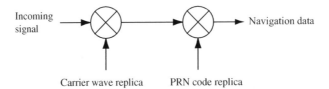

Incoming signal → ⊗ → ⊗ → Navigation data

Carrier wave replica   PRN code replica

FIGURE 7.1. Basic demodulation scheme. This scheme is used to demodulate the navigation message.

The output from the front end including filtering and downconversion can be described as

$$s^k(t) = \sqrt{2P_C}\,C^k(t)D^k(t)\cos(\omega_{IF}t) + \sqrt{2P_{PL1}}\,P^k(t)D^k(t)\sin(\omega_{IF}t), \quad (7.2)$$

where ω_{IF} is the intermediate frequency to which the front end has downconverted the carrier frequency. Equation (7.2) describes the output of the front end from one satellite.

This signal is then sampled by the A/D converter. Because of the narrow bandpass filter around the C/A code, the P code is distorted. In this way the last term in Equation (7.2) is filtered out and cannot be demodulated and is in the following described as noise $e(n)$. The signal from satellite k after the A/D conversion can be described as

$$s^k(n) = C^k(n)D^k(n)\cos(\omega_{IF}n) + e(n) \qquad (7.3)$$

with n in units of $1/f_s$ s; n indicates that the signal is discrete in time.

To obtain the navigation data $D^k(n)$ from the above signal, the signal has to be converted down to baseband. The carrier removal is done by multiplying the input signal with a replica of the carrier as shown in Figure 7.1. If the carrier replica is identical to the incoming carrier in both frequency and phase, the product of both is

$$s^k(n)\cos(\omega_{IF}n) = C^k(n)D^k(n)\cos(\omega_{IF}n)\cos(\omega_{IF}n)$$
$$= -\tfrac{1}{2}C^k(n)D^k(n) - \tfrac{1}{2}\cos(2\omega_{IF}n)C^k(n)D^k(n), \qquad (7.4)$$

where the first term is the navigation message multiplied with the PRN code and the second term is a carrier with the double intermediate frequency. The latter part of the signal can be removed by applying a lowpass filter. The signal after the lowpass filter is

$$\tfrac{1}{2}C^k(n)D^k(n). \qquad (7.5)$$

The next step is to remove the code $C^k(n)$ from the signal. This is done by correlating the signal with a local code replica. If the code replica is exactly the same as the code in the signal, the output of the correlation is

$$\sum_{n=0}^{N-1} C^k(n)C^k(n)D^k(n) = ND^k(n), \qquad (7.6)$$

where $ND^k(n)$ is the navigation message multiplied by the amount of points in the signal N.

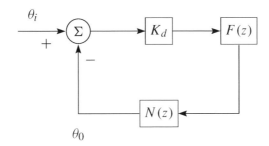

FIGURE 7.2. Linearized digital second-order PLL model.

The above description of the demodulation is only for a signal with one satellite. This is done to reduce the complexity of the equations and to give a simpler idea of the demodulation scheme. In the real signal there is a signal contribution from each visible satellite resulting in larger noise terms in the equations; see Haykin (2000), page 95.

In the demodulation scheme seen in Figure 7.1, two local signal replicas are required. To produce the exact replica some kind of feedback is needed. The feedback loop to produce the carrier replica is referred to as the carrier tracking loop, and the feedback loop to produce the exact code replica is referred to as the code tracking loop.

7.3   Second-Order PLL

Both the carrier tracking (Costas loop) and code tracking [delay lock loop (DLL)] have an analytic linear phase lock loop model that can be used to predict performance. This linear model has been derived by Ziemer & Peterson (1985) and is an extremely powerful tool to predict the performance of the tracking loop. Another excellent reference, once the fundamental models for Costas and DLL have been derived, for linear phase lock loop and its parameters and performance is by Best (2003).

Extending the linear PLL model has been derived by Chung et al. (1993). This approach will be followed by the implementation of both the Costas loop and DLL, yet the linear model references earlier can still be the basis of performance prediction and analysis.

The second-order PLL system contains a first-order filter and a voltage controlled oscillator (VCO). Note that the transfer function of an analog loop filter and a VCO are

$$F(s) = \frac{1}{s} \frac{\tau_2 s + 1}{\tau_1}, \tag{7.7}$$

$$N(s) = \frac{K_o}{s}, \tag{7.8}$$

where $F(s)$ and $N(s)$ are the transfer functions of the filter and NCO, respectively. K_o is the NCO gain. The transfer function of a linearized analog PLL is [refer to

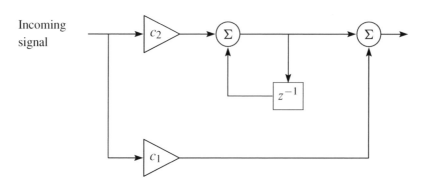

FIGURE 7.3. Second-order phase lock loop filter.

Chung et al. (1993)]

$$H(s) = \frac{K_d F(s) N(s)}{1 + K_d F(s) N(s)}, \tag{7.9}$$

where K_d is the gain of the phase discriminator. Substituting Equations (7.7) and (7.8) into the transfer function (7.9) yields

$$H(s) = \frac{2\zeta \omega_n s + \omega_n^2}{s^2 + s\zeta \omega_n s + \omega_n^2}, \tag{7.10}$$

where the natural frequency $\omega_n = \sqrt{(K_o K_d)/\tau_1}$, and the damping ratio $\zeta = (\tau_2 \omega_n)/2$. The above transfer functions are analog versions and to convert the transfer functions to digital form, the bilinear transformation is used on (7.10). This yields the following digital transfer functions for the PLL model:

$$H_1(z) = \frac{\left(4\zeta \omega_n T + (\omega_n T)^2\right) + 2(\omega_n T)^2 z^{-1} + \left((\omega_n T)^2 - 4\zeta \omega_n T\right) z^{-2}}{\left(4 + 4\zeta \omega_n T + (\omega_n T)^2\right) + \left(2(\omega_n T)^2 - 8\right) z^{-1} + \left(4 - 4\zeta \omega_n T + (\omega_n T)^2\right) z^{-2}}. \tag{7.11}$$

The linearized digital second-order PLL model is shown in Figure 7.2, where K_d is the discriminator gain, $F(z)$ is the transfer function of the filter, and $N(z)$ is the transfer function of NCO. The transfer functions for the digital filter and NCO are

$$F(z) = \frac{(C_1 + C_2) - C_1 z^{-1}}{1 - z^{-1}}, \tag{7.12}$$

$$N(z) = \frac{K_o z^{-1}}{1 - z - 1}, \tag{7.13}$$

where $F(z)$ is the transfer function of the filter and $N(z)$ is the transfer function of the NCO. Figure 7.3 shows the phase lock filter $F(z)$.

The goal is to find the coefficients C_1 and C_2 in the second-order PLL. This is done by comparing the transfer function for the digital PLL and the transfer function for the analog PLL. The transfer function for the digital version can be found as

$$H(z) = \frac{\theta_o(z)}{\theta_i(z)} = \frac{K_d F(z) N(z)}{1 + K_d F(z) N(z)}. \tag{7.14}$$

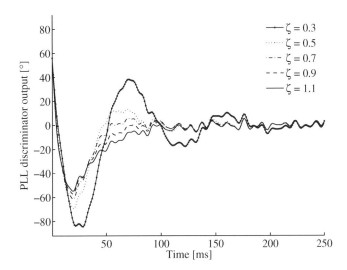

FIGURE 7.4. Phase error as function of different damping ratios ζ. A larger settling time results in a smaller overshoot of the phase.

By substituting (7.12) and (7.13) into (7.14), we obtain the following:

$$H_2(z) = \frac{K_o K_d (C_1 + C_2) z^{-1} - K_o K_d C_1 z^{-2}}{1 + \left(K_o K_d (C_1 + C_2) - 2\right) z^{-1} + (1 - K_o K_d C_1) z^{-2}}. \tag{7.15}$$

To find an equation for the two coefficients C_1 and C_2, (7.11) and (7.15) are compared. This yields the following two equations:

$$C_1 = \frac{1}{K_o K_d} \frac{8 \zeta \omega_n T}{4 + 4 \zeta \omega_n T + (\omega_n T)^2}, \tag{7.16}$$

$$C_2 = \frac{1}{K_o K_d} \frac{4 (\omega_n T)^2}{4 + 4 \zeta \omega_n T + (\omega_n T)^2}, \tag{7.17}$$

where $K_o K_d$ is the loop gain, ζ is the damping ratio, ω_n is the natural frequency, and T is the sampling time; see Chung et al. (1993).

The natural frequency can be found as

$$\omega_n = \frac{8 \zeta B_L}{4 \zeta^2 + 1}, \tag{7.18}$$

where B_L is the noise bandwidth in the loop; see Parkinson & Spilker Jr. (1996), volume 1, page 371.

The damping ratio and noise bandwidth are computed for a particular signal case. But in some cases an engineer would like to change these values for specific applications or implementations. Therefore, a more thorough explanation is given about these parameters.

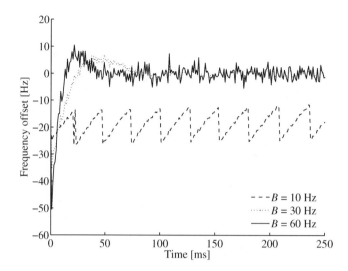

FIGURE 7.5. Frequency offsets from an acquired frequency offset 20 Hz and for PLL noise bandwidths of 10, 30, and 60 Hz. There are negative peaks in the first 2 ms due to transition phase in the loop filter.

7.3.1 Damping Ratio

The damping ratio controls how fast the filter reaches its settle point. The damping ratio also controls how much overshoot the filter can have. A smaller settling time results in a larger overshoot. This can be seen in Figure 7.4.

The choice of damping ratio is a compromise between overshoot and settling time. The damping is chosen to $\zeta = 0.7$ resulting in a filter that converges reasonably fast and does not make a high overshoot.

7.3.2 Noise Bandwidth

The second parameter in the PLL filter is the noise bandwidth B_L. The noise bandwidth controls the amount of noise allowed in the filter. This parameter can also, as the damping ratio, control the settling time. As the tracking loop starts to track a signal the start frequency is the frequency found by the acquisition algorithm. (This phase is sometimes called the pull-in phase in the literature. It is in this phase the filter is trying to converge to the correct frequency and phase.) The start frequency from the acquisition algorithm can be off by some Hz. The tracking loop is then going to lock onto the correct frequency. To see the impact of various noise bandwidths, a real GPS signal is used where the acquisition algorithm found a frequency that is about 21 Hz off. Figure 7.5 shows the offset from the start frequency for three different noise bandwidths.

From the figure, it follows that if the noise bandwidth is 60 Hz, the tracking loop immediately finds the correct frequency offset of about 21 Hz. It can also be seen that a lot of noise in the tracking frequency is allowed. In the second case, where the noise bandwidth is 30 Hz, the tracking loop also locks on the signal

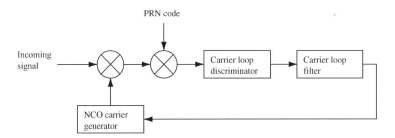

FIGURE 7.6. Basic GPS receiver tracking loop block diagram.

quite fast. It can be seen that the noise on the tracking frequency is much smaller than with a noise bandwidth of 60 Hz. In the third case where the noise bandwidth is 10 Hz, the tracking loop is not fast enough to reach the real frequency before a phase shift occurs. Therefore, it is not likely to converge to the proper value. A large noise bandwidth implies that the tracking loop quickly locks to the real frequency but has a relatively large frequency noise in the locked state. A smaller noise bandwidth implies that it can take some time before the tracking loop can be locked to the frequency, but after the lock the frequency is stable. Some implementations split the PLL into two filters, often called pull-in and tracking filters.

For land applications, a typical value for noise bandwidth is about 20 Hz.

7.4 Carrier Tracking

To demodulate the navigation data successfully an exact carrier wave replica has to be generated. To track a carrier wave signal, phase lock loops (PLL) or frequency lock loops (FLL) are often used.

Figure 7.6 shows a basic block diagram for a phase lock loop. The two first multiplications wipe off the carrier and the PRN code of the input signal. To wipe off the PRN code, the I_p output from the early–late code tracking loop described above is used. The loop discriminator block is used to find the phase error on the local carrier wave replica. The output of the discriminator, which is the phase error (or a function of the phase error), is then filtered and used as a feedback to the numerically controlled oscillator (NCO), which adjusts the frequency of the local carrier wave. In this way the local carrier wave could be an almost precise replica of the input signal carrier wave.

The problem with using an ordinary PLL is that it is sensitive to 180° phase shifts. Due to navigation bit transitions, a PLL used in a GPS receiver has to be insensitive to 180° phase shifts.

Figure 7.7 shows a Costas loop. One property of this loop is that it is insensitive for 180° phase shifts and hereby a Costas loop is insensitive for phase transitions due to navigation bits. This is the reason for using this carrier tracking loop in GPS receivers. The Costas loop in Figure 7.7 contains two multiplications. The first multiplication is the product between the input signal and the local carrier wave and the second multiplication is between a 90° phase-shifted carrier wave

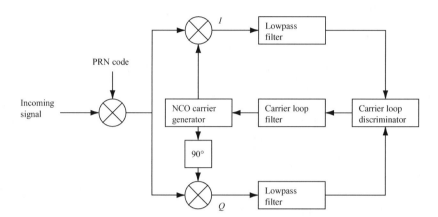

FIGURE 7.7. Costas loop used to track the carrier wave.

and the input signal. The *goal of the Costas loop is to try to keep all energy in the I (in-phase) arm*. To keep the energy in the I arm, some kind of feedback to the oscillator is needed. If it is assumed that the code replica in Figure 7.7 is perfectly aligned, the multiplication in the I arm yields the following sum:

$$D^k(n)\cos(\omega_{\text{IF}}\,n)\cos(\omega_{\text{IF}}\,n+\varphi) = \tfrac{1}{2}D^k(n)\cos(\varphi) + \tfrac{1}{2}D^k(n)\cos(2\omega_{\text{IF}}\,n+\varphi),$$
(7.19)

where φ is the phase difference between the phase of the input signal and the phase of the local replica of the carrier phase. The multiplication in the quadrature arm gives the following:

$$D^k(n)\cos(\omega_{\text{IF}}\,n)\sin(\omega_{\text{IF}}\,n+\varphi) = \tfrac{1}{2}D^k(n)\sin(\varphi) + \tfrac{1}{2}D^k(n)\sin(2\omega_{\text{IF}}\,n+\varphi).$$
(7.20)

If the two signals are lowpass filtered after the multiplication, the two terms with the double intermediate frequency are eliminated and the following two signals remain:

$$I^k = \tfrac{1}{2}D^k(n)\cos(\varphi),$$
(7.21)

$$Q^k = \tfrac{1}{2}D^k(n)\sin(\varphi).$$
(7.22)

To find a term to feed back to the carrier phase oscillator, it can be seen that the phase error of the local carrier phase replica can be found as

$$\frac{Q^k}{I^k} = \frac{\tfrac{1}{2}D^k(n)\sin(\varphi)}{\tfrac{1}{2}D^k(n)\cos(\varphi)} = \tan(\varphi),$$
(7.23)

$$\varphi = \tan^{-1}\left(\frac{Q^k}{I^k}\right).$$
(7.24)

From Equation (7.24), it can be seen that the phase error is minimized when the correlation in the quadrature-phase arm is zero and the correlation value in the in-phase arm is maximum. The arctan discriminator in Equation (7.24) is the most precise of the Costas discriminators, but it is also the most time-consuming. Table 7.1 describes other possible Costas discriminators.

TABLE 7.1. Various types of Costas phase lock loop discriminators

Discriminator	Description
$D = \text{sign}(I^k)Q^k$	The output of the discriminator is proportional to $\sin(\varphi)$.
$D = I^k Q^k$	The discriminator output is proportional to $\sin(2\varphi)$.
$D = \tan^{-1}\left(\frac{Q^k}{I^k}\right)$	The discriminator output is the phase error.

Figure 7.8 shows the responses corresponding to the different discriminators. The phase discriminator outputs in this figure are computed using expressions in Table 7.1 for all possible phase errors. In the same figure it can be seen that the discriminator outputs are zero when the real phase error is 0 and $\pm 180°$. This is why the Costas loop is insensitive to the $180°$ phase shifts in case of a navigation bit transition.

The behavior of the Costas loop when a $180°$ phase shift occurs is more clearly illustrated in Figure 7.9. In this figure the vector sum of I^k and Q^k is shown as the vector in the coordinate system. If the local carrier wave were in phase with the input signal, the vector would be aligned to the I-axis, but in the present case a small phase error is illustrated. When the signal is tracked correctly the vector sum of I^k and Q^k tends to remain aligned with the I-axis. This property ensures that if a navigation bit transition occurs, the vector on the phasor diagram will flip $180°$ (showed by the dashed vector in the figure). If a navigation bit transition occurs, the Costas loop will still track the signal and nothing will happen. This property does make Costas loop the commonly chosen phase lock loop in GPS receivers; see Kaplan & Hegarty (2006), pages 166–170.

The output of the phase discriminator is filtered to predict and estimate any relative motion of the satellite and to estimate the Doppler frequency.

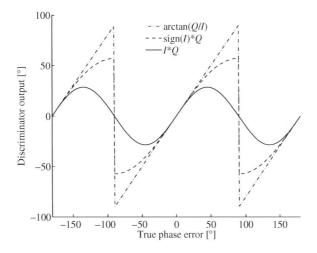

FIGURE 7.8. Comparison between the common Costas phase lock loop discriminator responses.

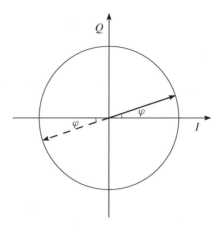

FIGURE 7.9. Phasor diagram showing the phase error between the phase of the input carrier wave and the phase of the local carrier wave replica.

7.5   Code Tracking

The goal for a code tracking loop is to keep track of the code phase of a specific code in the signal. The output of such a code tracking loop is a perfectly aligned replica of the code. The code tracking loop in the GPS receiver is a delay lock loop (DLL) called an early–late tracking loop. The idea behind the DLL is to correlate the input signal with three replicas of the code seen in Figure 7.10.

The first step in Figure 7.10 is converting the C/A code to baseband, by multiplying the incoming signal with a perfectly aligned local replica of the carrier wave. Afterwards the signal is multiplied with three code replicas. The three replicas are nominally generated with a spacing of $\pm\frac{1}{2}$ chip. After this second multiplication, the three outputs are integrated and dumped. The output of these integrations is a numerical value indicating how much the specific code replica correlates with the code in the incoming signal.

The three correlation outputs I_E, I_P, and I_L are then compared to see which one provides the highest correlation. Figure 7.11 shows an example of code tracking. In Figure 7.11a the late code has the highest correlation, so the code phase must

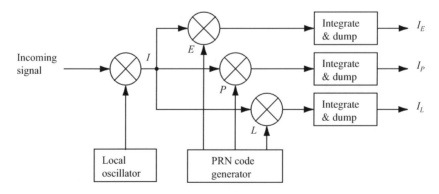

FIGURE 7.10. Basic code tracking loop block diagram.

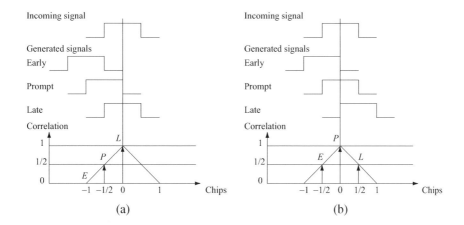

FIGURE 7.11. Code tracking. Three local codes are generated and correlated with the incoming signal. (a) The late replica has the highest correlation so the code phase must be decreased, i.e., the code sequence must be delayed. (b) The prompt code has the highest correlation, and the early and late have similar correlation. The loop is perfectly tuned in.

be decreased. In Figure 7.11b the highest peak is located at the prompt replica, and the early and late replicas have equal correlation. In this case, the code phase is properly tracked.

The DLL with three correlators as in Figure 7.10 is optimal when the local carrier wave is locked in phase and frequency. But when there is a phase error on the local carrier wave, the signal will be more noisy, making it more difficult for the DLL to keep lock on the code. So instead the DLL in a GPS receiver is often designed as in Figure 7.12.

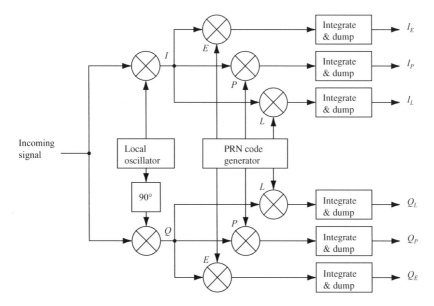

FIGURE 7.12. DLL block diagram with six correlators.

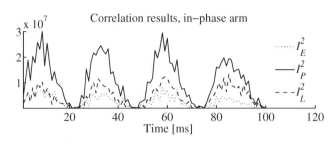

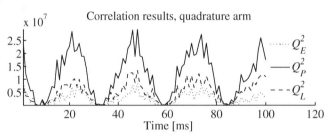

FIGURE 7.13. Output of the six correlators in the in-phase and quadrature arms of the tracking loop. Acquisition frequency offset is 20 Hz and PLL noise bandwidth is 15 Hz (for demonstration purpose).

The design in Figure 7.12 has the advantage that it is independent of the phase on the local carrier wave. If the local carrier wave is in phase with the input signal, all the energy will be in the in-phase arm. But if the local carrier phase drifts compared to the input signal, the energy will switch between the in-phase and the quadrature arm. For demonstration purposes, Figure 7.13 shows such a situation

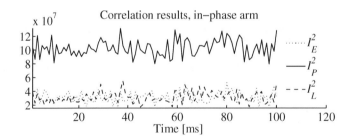

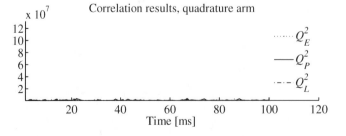

FIGURE 7.14. Output of the six correlators in the in-phase and quadrature arms of the tracking loop. The local carrier wave is in phase with the input signal.

TABLE 7.2. Various types of delay lock loop discriminators and a description of them

Type	Discriminator D	Characteristics
Coherent	$I_E - I_L$	Simplest of all discriminators. Does not require the Q branch but requires a good carrier tracking loop for optimal functionality.
Noncoherent	$(I_E^2 + Q_E^2) - (I_L^2 + Q_L^2)$	Early minus late power. The discriminator response is nearly the same as the coherent discriminator inside $\pm\frac{1}{2}$ chip.
	$\dfrac{(I_E^2 + Q_E^2) - (I_L^2 + Q_L^2)}{(I_E^2 + Q_E^2) + (I_L^2 + Q_L^2)}$	Normalized early minus late power. The discriminator has a great property when the chip error is larger than a $\frac{1}{2}$ chip; this will help the DLL to keep track in noisy signals.
	$I_P(I_E - I_L) + Q_P(Q_E - Q_L)$	Dot product. This is the only DLL discriminator that uses all six correlator outputs.

where the phase of the carrier replica drifts compared to the phase of the incoming signal. The upper plot shows the output of the three correlators in the in-phase arm, and the lower plot shows the correlation output in the quadrature arm of the DLL with six correlators. This situation is a result of different frequencies for the signal and the replica; it results in a constantly changing phase difference (misalignment). There are a few reasons why this can happen, for example, the PLL could be not in a lock state.

Figure 7.14 shows a case when the PLL is in a lock state. Because of the precise carrier replica from the PLL, it is seen in Figure 7.14 that the correlators are constant over time. This would not be the case if the carrier replica is not adjusted to match the frequency and phase of the incoming signal.

If the code tracking loop performance has to be independent of the performance of the phase lock loop, the tracking loop has to use both the in-phase and quadrature arms to track the code.

The DLL now needs a feedback to the PRN code generators if the code phase has to be adjusted. Some common DLL discriminators used for feedback are listed in Table 7.2.

The table shows one coherent and three noncoherent discriminators. The requirements of a DLL discriminator is dependent on the type of application and the noise in the signal. The discriminator function responses are shown in Figure 7.15.

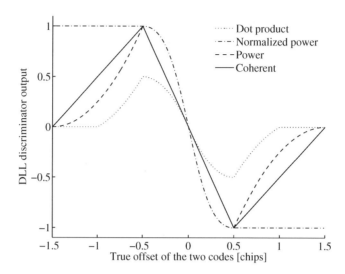

FIGURE 7.15. Comparison between the common DLL discriminator responses.

Figure 7.15 shows the coherent discriminator and three noncoherent discriminators using a standard correlator. The figure is produced from ideal ACFs, and the space between the early, prompt, and late is $\pm\frac{1}{2}$ chip. The space between the early, prompt, and late codes determines the noise bandwidth in the delay lock loop. If the discriminator spacing is larger than $\frac{1}{2}$ chip, the DLL would be able to handle wider dynamics and be more noise robust; on the other hand, a DLL with a smaller spacing would be more precise. In a modern GPS receiver the discriminator spacing can be adjusted while the receiver is tracking the signal. The advantage from this is that if the signal-to-noise ratio suddenly decreases, the receiver uses a wider spacing in the correlators to be able to handle a more noisy signal, and hereby a possible code lock loss could be avoided; see Kaplan & Hegarty (2006), page 175.

The implemented tracking loop discriminator is the normalized early minus late power. This discriminator is described as

$$D = \frac{(I_E^2 + Q_E^2) - (I_L^2 + Q_L^2)}{(I_E^2 + Q_E^2) + (I_L^2 + Q_L^2)}, \qquad (7.25)$$

where I_E, Q_E, I_L, and Q_L are output from four of the six correlators shown in Figure 7.13. The normalized early minus late power discriminator is chosen because it is independent of the performance of the PLL as it uses both the in-phase and quadrature arms. The normalization of the discriminator causes that the discriminator can be used with signals with different signal-to-noise ratios and different signal strengths.

The tracking loop generates three local code replicas. In this section, the chip space between the early and prompt replicas is half a chip.

As was described, the DLL can be modeled as a linear PLL and thus the performance of the loop can be predicted based on this model. In other words the loop filter design is the same, just the parameter values are different.

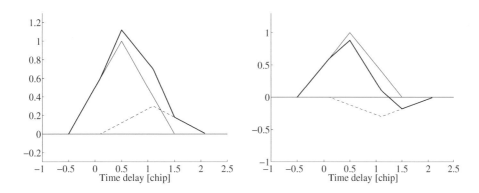

FIGURE 7.16. Constructive and destructive multipath interference. The solid thin line represents the original triangular correlation function; the dashed line the multipath correlation function, and the third line is the sum of the two.

7.6   Multipath

Of the multiple error sources associated with GNSS signal processing, multipath directly impacts the code tracking performance. With the prior description of how code tracking is implemented, it makes sense to investigate how multipath impacts the code tracking loop.

The signal observed at the receiver is a distorted version of the one transmitted. One distortion effect is called multipath propagation or, for short, *multipath.*

If the receiver can directly see the satellite, a part of the received signal has propagated via the direct path from the satellite to the receiver. This signal component is delayed according to the distance between the satellite transmitter and the receiver. The signal component propagated via the direct path usually is by far the strongest part of the received signal.

In addition to the direct signal, the receiver may observe other signals propagating via other and longer paths. This can happen if the radio wave reaches the receiver after interaction with one or more objects/obstacles in the environment. Different kinds of interactions between radio waves and objects exist, but in the context of GNSS it suffices to think of an interaction as a reflection altering the direction of propagation, amplitude, polarity, and phase of the radio wave.

First we consider the simple case where the transmitted signal reaches the receiver via two paths. Thus, the received signal consists of two components: a direct component and a signal component reaching the receiver via a reflection on a nearby building. In this case the reflected signal component is a delayed, phase-shifted, and attenuated version of the line-of-sight signal component. Due to the change in phase and delay, the two signal components interfere. In case the reflected and the direct signals are in phase, the amplitude of the sum is larger than the amplitude of each of the components. This we call *constructive interference*; see Figure 7.16. On the other hand, if the direct and reflected signals are out of phase, the amplitude of the sum signal decreases and we talk about *destructive interference*. In case the relative phase between the direct and reflected signals rapidly changes the amplitude of the sum signal, we say that the received signal is *fading*. A signal component reaching the receiver via a different path than the direct path is called a *multipath component*.

In general, the received signal $x(t)$ is composed of the direct signal and $M - 1$ multipath components. Let $A_i(t)$ denote the amplitude of the ith multipath component, let D denote the navigation message, let C denote the code, let τ denote the multipath error, let the frequency change be v_i, the phase offset be φ_i, and, finally, add a noise term $n(t)$; then the signal can be described as

$$x(t) = \sum_{i=1}^{M} A_i(t)D\big(t - \tau_i(t)\big)C\big(t - \tau_i(t)\big)\cos\big(2\pi(f_0 + v_i(t))t + \varphi_i(t)\big) + n(t).$$

To simplify the discussion we consider a two-path scenario ($M = 2$) and make the following assumptions:

$$A_1(t) = A_1; \qquad\qquad A_2(t) = A_2;$$
$$\tau_1(t) = \tau_1; \qquad\qquad \tau_2(t) = \tau_2;$$
$$v_1(t) = v_1; \qquad\qquad v_2(t) = v_2;$$
$$n(t) = 0.$$

In other words we have assumed that the parameters (amplitudes, delays, and Doppler shifts) are constant over the time period we consider. In this case, $x(t)$ can be reduced to

$$x(t) = A_1 D(t - \tau_1)C(t - \tau_1)\cos\big(2\pi(f_0 + v_1)t + \varphi_1\big)$$
$$+ A_2 D(t - \tau_2)C(t - \tau_2)\cos\big(2\pi(f_0 + v_2)t + \varphi_2\big). \quad (7.26)$$

We now exploit $\cos(\varphi) = \Re\big(\exp(j\varphi)\big)$ to rewrite (7.26) as

$$x(t) = \Re\Big\{ A_1 D(t - \tau_1)C(t - \tau_1)\exp\big[j\big(2\pi(f_0 + v_1)t + \varphi_1\big)\big]$$
$$+ A_2 D(t - \tau_2)C(t - \tau_2)\exp\big[j\big(2\pi(f_0 + v_2)t + \varphi_2\big)\big]\Big\}$$
$$= \Re\Big\{ A_1 D(t - \tau_1)C(t - \tau_1)\exp\big[j\big(2\pi(f_0 + v_1)t + \varphi_1\big)\big]$$
$$+ A_2 D(t - \tau_2)C(t - \tau_2)\exp\big[j\big(2\pi(v_2 - v_1)t + (\varphi_2 - \varphi_1)\big)\big]$$
$$\times \exp\big[j\big(2\pi(f_0 + v_1)t + \varphi_1\big)\big]\Big\}$$
$$= \Re\Big\{ \Big(A_1 D(t - \tau_1)C(t - \tau_1)$$
$$+ A_2 D(t - \tau_2)C(t - \tau_2)\exp\big[j\big(2\pi(v_2 - v_1)t + (\varphi_2 - \varphi_1)\big)\big]\Big)$$
$$\times \exp\big[j\big(2\pi(f_0 + v_1)t + \varphi_1\big)\big]\Big\}.$$

Defining the instantaneous phase difference between the two signal components $\psi(t) = 2\pi(v_2 - v_1)t + (\varphi_2 - \varphi_1)$, the output of the integrators in the DLL can be

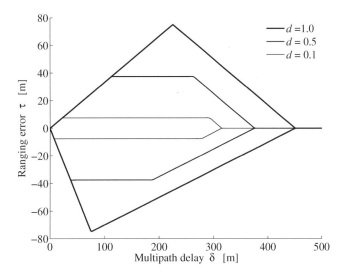

FIGURE 7.17. Multipath error envelope for noncoherent early/late detector for C/A code, $\alpha = 0.5$. The positive multipath error corresponds to constructive interference while negative multipath error corresponds to destructive interference.

approximated as

$$y_\pm(t) \approx A_1 R_C(\tau_1 - \hat{\tau} \pm \delta) D(t - \tau_1) + A_2 R_C(\tau_2 - \hat{\tau} \pm \delta) \exp\big(j\psi(t)\big) D(t - \tau_2)$$
$$\approx \Big[A_1 R_C(\tau_1 - \hat{\tau} \pm \delta) + A_2 \exp\big(j\psi(t)\big) R_C(\tau_2 - \hat{\tau} \pm \delta) \Big] D(t - \tau_1).$$

With $\epsilon = \tau_1 - \hat{\tau}$, the envelope of $y_\pm(t)$ is

$$|y_\pm(t)| = \Big| A_1 R_C(\epsilon \pm \delta) + A_2 \exp\big(j\psi(t)\big) R_C\big(\epsilon + (\tau_2 - \tau_1) \pm \delta\big) \Big|. \quad (7.27)$$

The output of the envelope discriminator is

$$\epsilon(t) = |y_-(t)| - |y_+(t)|. \quad (7.28)$$

As mentioned before, the amplitude of the sum signal depends on the relative phase $\psi(t)$ between the direct and the reflected signal components. We consider the two special cases of constructive and destructive interference:

Constructive interference In this case $\exp\big(j\psi(t)\big) = 1$ and

$$|y_\pm(t)| = A_1 R_C(\epsilon \pm \delta) + A_2 R_C(\epsilon + \tau_2 - \tau_1 \pm \delta). \quad (7.29)$$

Destructive interference Here, $\exp\big(j\psi(t)\big) = -1$ and, therefore,

$$|y_\pm(t)| = A_1 R_C(\epsilon \pm \delta) - A_2 R_C(\epsilon + \tau_2 - \tau_1 \pm \delta). \quad (7.30)$$

If the receiver is moving, it is very likely that $v_1 \neq v_2$ and the relative phase changes over time. Therefore, the envelope of the received signal will fluctuate

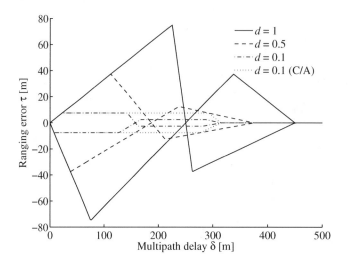

FIGURE 7.18. Multipath error envelope for noncoherent early/late detector for BOC(1,1). Negative ranging error corresponds to destructive interference.

over time as the two signal components interfere. As a measure of the magnitude of the fluctuations we define the *relative amplitude* as

$$\alpha = \frac{A_2}{A_1}.$$

In practice, the amplitude of the direct component is larger than the reflected component and thus α is somewhat less than one.

Equation (7.29) is graphed as the upper half of Figure 7.17 for $\alpha = 0.5$ and for three values of the correlator spacing $d = 0.1, 0.5,$ and 1.0. The upper part of the figure corresponds to constructive interference while the lower part corresponds to destructive interference; see Equation (7.30). All actual multipath errors lie within the combined envelope!

In Figure 7.17 the initial slope is a function of multipath amplitude and delay δ only. It is independent of correlator and PRN chipping rate f_c. We recall the C/A-code pseudorange multipath error can approach 147 m, theoretically. However, errors of 10 m or less are far more common. Large errors can be encountered in urban environments.

Our investigation assumes infinite GPS signal bandwidth. Bandwidths of 10–20 MHz yield results that are similar to those for the infinite bandwidth case. For short-delay multipath (small δ), the finite bandwidth effects are much less significant.

In conclusion: Multipath propagation deforms the ideal correlation peak because the received signal is a sum of signal components. The multipath components arrive later at the receiver and contribute additional correlation peaks. Thus, the early–late correlator samples may not be centered on the true arrival time of the direct path.

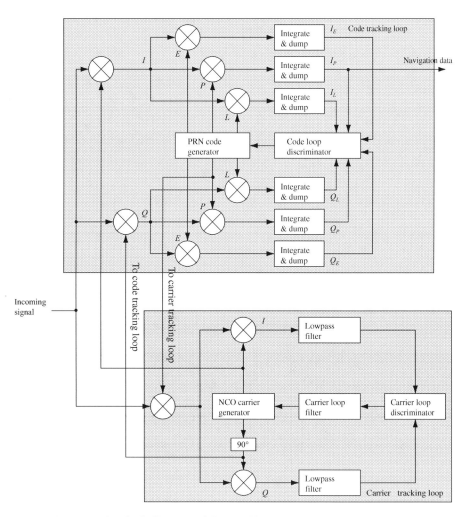

FIGURE 7.19. Block diagram of the combined DLL and PLL tracking loops.

We note that in MATLAB the ACF can be coded as

R = (1 − abs(tau)) ∗ heaviside(1 − abs(tau));

Next we determine the multipath error envelope for Galileo BOC(1,1). The result is plotted in Figure 7.18. Comparing Figures 7.17 and 7.18 for correlator spacing $d = 0.1$ we see that the C/A code multipath error envelop is sensitive for multipath signals with a relative path delay up to 300 m. The resulting range errors τ are between ±4 m. For BOC(1,1) the corresponding value is 150 m and range errors of ±4 m. This demonstrates that BOC(1,1) signals are better to handle multipath signals than C/A code signals!

The multipath error envelopes are computed from equations like (7.29) and (7.30). Figures 7.17 and 7.18 are plotted by using the powerful ezplot command. With correlator spacing $d = 0.1$ and multipath delays δ in the range 0–157 m

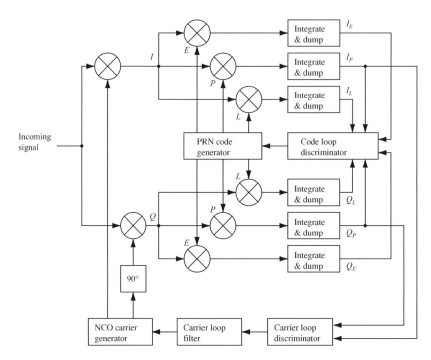

FIGURE 7.20. The block diagram of a complete tracking channel on the GPS receiver.

we obtain ranging errors τ in the interval ± 7.5 m; for δ in the range 157–317 m the delay τ is within ± 2.5 m. Obviously the delay range τ strongly depends on the correlator spacing d; some papers even mention $d = 1/25$. See also Winkel (2005).

7.7   Complete Tracking Block

In the previous sections, the code tracking loop and the carrier tracking loop are described in detail. The following describes how the code tracking loop and the carrier tracking loop can be joined to minimize the computational load.

Figure 7.19 shows the code tracking loop and the carrier tracking loop combined. It can be seen from the figure that the PRN code replica used to wipe off the PRN code in the carrier tracking loop is coming from the code tracking loop. It can also be seen that the two local carrier replicas used to wipe off the carrier wave in the code tracking loop are coming from the carrier tracking loop. The block diagram in Figure 7.19 contains 11 multiplications. These multiplications are the most time-consuming operations on the block diagram.

Figure 7.20 shows an optimized version of the combined tracking loops. Here the I and Q inputs to the phase discriminator are the I_p and Q_p correlation from the code tracking loop. In this way the three multiplications in the Costas loop are eliminated, and hereby the computation time is reduced.

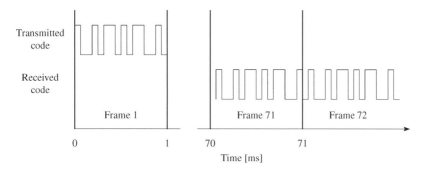

FIGURE 7.21. The delay between the time of transmission at the satellite and time of reception at the receiver.

7.8   Pseudorange Computations

Precise estimation of the pseudorange from a satellite to the receiver is crucial for a modern C/A code GPS receiver. The relationship between the standard deviation of the observations and that of the coordinates of the receiver position (see page 131) is

$$\sigma_{\text{pos}} = \sqrt{\sigma_e^2 + \sigma_n^2 + \sigma_u^2} = \text{PDOP } \sigma_0, \qquad (7.31)$$

where σ_{pos} is the standard deviation of the receiver position, σ_0 is the standard deviation of unit weight. PDOP is the position dilution of precision, which depends on the geometry of the satellite constellation. Optimal accuracy of the position is obtained when the standard deviation of unit weight is as small as possible.

A pseudorange measurement is computed as the travel time from the satellite to the receiver multiplied by the speed of light in vacuum. The receiver has to estimate exactly when the start of a frame arrives at the receiver. This is done by adding the code phase to the time when the frame entered the receiver.

In Figure 7.21, the satellite transmits the start of the C/A code at $t = 0$ ms. This signal is received by the receiver approximately 70 ms after it is transmitted from the satellite. A range from the satellite to the receiver of 21,000 km corresponds to a travel time of 70 ms. As described before, the receiver is using block processing. That is, to calculate an accurate pseudorange and hereby an accurate position, the exact start of the C/A code in frame 71 in Figure 7.21 has to be found.

Figure 7.22 shows the first 700 samples of frame 71 in detail. The receiver has a time tag for the start of the frame. The problem is then to determine exactly where the start of the code is in the frame of data. In Figure 7.22, the start of the C/A code is at sample number 605, which is the correct code start in the simulated data used in this section.

Since the sampling frequency of the reference data set is at 38.192 MHz, each sample corresponds to

$$\Delta p = \frac{c}{38.192\,\text{MHz}} = 7.86\,\text{m}.$$

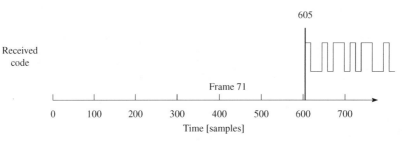

FIGURE 7.22. Close-up of frame 71 from Figure 7.21. In this case, the beginning of the C/A code is at sample 605 of the total of 38,192.

Since the prompt code is precisely aligned with the incoming signal to the nearest sample, the maximum error as a result of the discrete samples will be half that or < 5 m, which is sufficient for C/A code-type signals. If the sampling frequency is lower, a higher precision is required. Then it is possible to use the residual code phase at the end of each ms period to further refine that accuracy.

8
Data Processing for Positioning

8.1 Navigation Data Recovery

The output from the tracking loop is the value of the in-phase arm of the tracking block truncated to the values 1 and -1. Theoretically we could obtain a bit value every ms. However, we deal with noisy and weak signals, so a mean value for 20 ms is computed and truncated to -1 or 1. One navigation bit durates 20 ms.

The bit rate of the navigation data is 50 bps. The sample rate of the output from the tracking block is 1000 sps corresponding to a value each ms. Before the navigation data can be decoded, the signal from the tracking block must be converted from 1000 sps to 50 bps. That is, 20 consecutive values must be replaced by only 1. This conversion procedure is referred to as *bit synchronization*.

8.1.1 Finding the Bit Transition Time and the Bit Values

The first task of the bit synchronization procedure is to find the time in a sequence where bit transitions occur. First, a zero crossing is detected. A zero crossing is where the output changes from 1 to -1, or vice versa. When a zero crossing is located, the time of a bit transition is located. When the time of one bit transition is known, it is possible to find all bit transition times. These are located 20 ms apart beginning from the first detected bit transition. Figure 8.1 shows how all bit transition times are found in a 200-ms sequence. The bit transition times are marked by the arrows.

When the bit transition times are located, the 1000-bps signal must be converted to a bit rate of 50 bps. To do this, 20 samples must be replaced by only 1 value.

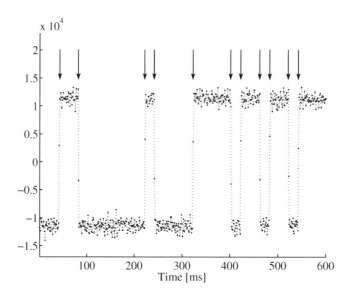

FIGURE 8.1. The figure shows output from the tracking block as dots. The dotted lines mark the bit transitions 20 ms apart. The actual signal is very strong, as a weaker one will have dots closer to zero.

8.2   Navigation Data Decoding

The navigation data *encoding* follows a scheme defined in the GPS Interface Control Document, ICD-GPS-200 (1991). The encoding scheme for Galileo is not available at the moment of writing.

When the GNSS navigation bits have been obtained through the bit synchronization, they must be *decoded*. The GPS ephemeris parameters are described below, while the tentative Galileo scheme is described in Section 3.4.

8.2.1   Location of Preamble

The first problem in the GPS navigation data decoding is to determine the location of the beginning of a subframe. The beginning of a subframe is marked by an 8-bit-long preamble. The pattern of the preamble is 10001011. Because of the Costas loop's ability to track the signal with a 180° phase shift, this preamble can occur in an inverted version 01110100. Naturally, these two bit patterns can occur anywhere in the received data so an additional check must be carried out to authenticate the preamble. The authentication procedure checks if the same preamble is repeated every 6 s corresponding to the time between transmission of two consecutive subframes.

The preamble search is implemented through a correlation. The first input to the correlation function is the incoming sequence of navigation data bits. This sequence is represented with −1's and 1's. The second input to the correlation function is the 8-bit preamble also represented with −1's and 1's. When using values −1 and 1 instead of 0 and 1, the output of the correlation function is 8

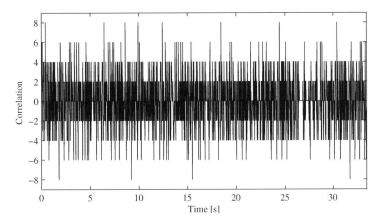

FIGURE 8.2. Correlation between 33 s of navigation data and the 8-bit preamble. The peaks indicate the location of the beginning of a subframe.

when the preamble is located and -8 when an inverted preamble is located. An example of the correlation between a navigation data sequence and the preamble can be seen in Figure 8.2.

As seen from Figure 8.2, the correlation function gives maximum correlation of 8 several times in this 33-s-long sequence. It should give only six maximum correlation values as the sequence should contain six subframes. In addition to the large number of maximum correlation values, it also gives a minimum correlation value of -8 several times. As mentioned earlier, this means that an inverted instance of the preamble has been located. The method for distinguishing which of the maximum correlation values that really is a beginning of a subframe includes the determination of the delay between consecutive maximum correlation values. Only if the delay between the maximum correlation values is exactly 6 s and the parity checks do not fail is the beginning of a subframe indicated.

When the correct preambles are located, the data from each subframe can be extracted. If the correlation shows that the preamble is inverted, the entire navigation sequence must be inverted.

Due to the Doppler effect the length of the navigation bit can deviate from the exact value of 20 ms. Over a short time this length difference even may accumulate to a significant value. Therefore, a better solution is to look for a preamble in the original 1000-sps output from the tracking. The algorithm remains the same, but each bit in the reference preamble pattern is converted to 20 values (samples). Now the correlation peak will have a maximum value of $8 \times 20 = 160$ instead of 8. Simultaneously, this modified algorithm also finds bit transition time.

8.2.2 Extracting the Navigation Data

Every correct preamble marks the beginning of a navigation data subframe. Each of the subframes contains 300 bits divided into 10 30-bit words. The structure of the first two words of a subframe is shown in Figure 8.3.

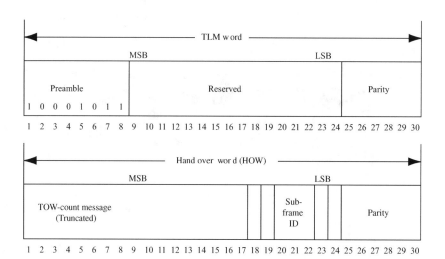

FIGURE 8.3. The first two words of each subframe. These words are referred to as the telemetry (TLM) word and the hand over word (HOW).

Parity Check   Besides 24 bits of data, every 30-bit word contains a 6-bit parity. The parity is used to check for misinterpreted bits in the navigation data. The parity is computed through the equations in Table 8.1. Here $\oplus$ denotes the modulo-2 or exclusive OR operation.

$D_1 - D_{24}$ are the 24 data bits in a word, while $D_{25} - D_{30}$ are the 6 parity bits in a word. The two bits denoted D_{29}^* and D_{30}^* are the last two parity bits from the previous word. When the navigation data are received, a parity check must be performed to test if the received data are interpreted correctly.

Time of Transmission   When the parity check has been performed successfully, the contents of the navigation data sequence can be decoded. The decoding is following the scheme from, ICD-GPS-200 (1991), which gives details of every word similar to what is shown in Figure 8.3. The first important issue is to determine the time when the current subframe was transmitted from the GPS satellite.

TABLE 8.1. Parity encoding equations

$$
\begin{aligned}
D_1 &= d_1 \oplus D_{30}^* \\
D_2 &= d_2 \oplus D_{30}^* \\
D_3 &= d_3 \oplus D_{30}^* \\
&\;\;\vdots \\
D_{24} &= d_{24} \oplus D_{30}^* \\
D_{25} &= D_{29}^* \oplus d_1 \oplus d_2 \oplus d_3 \oplus d_5 \oplus d_6 \oplus d_{10} \oplus d_{11} \oplus d_{12} \oplus d_{13} \oplus d_{14} \oplus d_{17} \oplus d_{18} \oplus d_{20} \oplus d_{23} \\
D_{26} &= D_{30}^* \oplus d_2 \oplus d_3 \oplus d_4 \oplus d_6 \oplus d_7 \oplus d_{11} \oplus d_{12} \oplus d_{13} \oplus d_{14} \oplus d_{15} \oplus d_{18} \oplus d_{19} \oplus d_{21} \oplus d_{24} \\
D_{27} &= D_{29}^* \oplus d_1 \oplus d_3 \oplus d_4 \oplus d_5 \oplus d_7 \oplus d_8 \oplus d_{12} \oplus d_{13} \oplus d_{14} \oplus d_{15} \oplus d_{16} \oplus d_{19} \oplus d_{20} \oplus d_{22} \\
D_{28} &= D_{30}^* \oplus d_2 \oplus d_4 \oplus d_5 \oplus d_6 \oplus d_8 \oplus d_9 \oplus d_{13} \oplus d_{14} \oplus d_{15} \oplus d_{16} \oplus d_{17} \oplus d_{20} \oplus d_{21} \oplus d_{23} \\
D_{29} &= D_{30}^* \oplus d_1 \oplus d_3 \oplus d_5 \oplus d_6 \oplus d_7 \oplus d_9 \oplus d_{10} \oplus d_{14} \oplus d_{15} \oplus d_{16} \oplus d_{17} \oplus d_{18} \oplus d_{21} \oplus d_{22} \oplus d_{24} \\
D_{30} &= D_{29}^* \oplus d_3 \oplus d_5 \oplus d_6 \oplus d_8 \oplus d_9 \oplus d_{10} \oplus d_{11} \oplus d_{13} \oplus d_{15} \oplus d_{19} \oplus d_{22} \oplus d_{23} \oplus d_{24}
\end{aligned}
$$

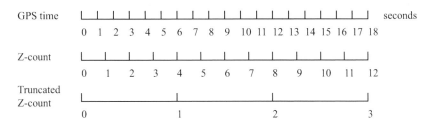

FIGURE 8.4. The relation between GPS time, Z-count, and the truncated Z-count in the HOW of the navigation data.

The second word of every subframe is the so-called HOW that includes a truncated version of the TOW. This number is referred to as the Z-count. The Z-count is the number of seconds passed since the last GPS week rollover in units of 1.5 s. The rollover happens at midnight between Saturday and Sunday. The maximum value of the Z-count is 403,199 as one week contains 604,800 s and 604,800 s/1.5 = 403,200 s. The Z-count value in the HOW is a truncated version containing only the 17 most significant bits (MSB). This truncation makes the Z-count increase in 6-s steps corresponding to the time between transmission of two consecutive navigation subframes. Figure 8.4 shows the relation between the three time measures: GPS time, Z-count, and truncated Z-count.

TABLE 8.2. Decoding scheme for GPS ephemeris parameters in the navigation data. n^* means that the current n bits should be decoded using the two's-complement.

Parameter	No. of bits	Scale factor (LSB)	Unit
IODE	8		
C_{rs}	16*	2^{-5}	m
Δn	16*	2^{-43}	semicircle/s
μ_0	32*	2^{-31}	semicircle
C_{uc}	16*	2^{-29}	radian
e	32	2^{-33}	dimensionless
C_{us}	16*	2^{-29}	radian
$\sqrt{a}$	32	2^{-19}	$m^{1/2}$
t_{oe}	16	2^4	s
C_{ic}	16*	2^{-29}	radian
Ω_0	32*	2^{-31}	semicircle
C_{is}	16*	2^{-29}	radian
i_0	32*	2^{-31}	semicircle
C_{rc}	16*	2^{-5}	m
ω	32*	2^{-31}	semicircle
$\dot{\Omega}$	24*	2^{-43}	semicircle/s
$\dot{i}$	14*	2^{-43}	semicircle/s

TABLE 8.3. Ephemeris parameters

IODE	issue of data, ephemeris
Δn	mean motion correction
μ_0	mean anomaly at t_{oe}
e	eccentricity
$\sqrt{a}$	square root of semi-major axis
t_{oe}	reference epoch of ephemeris
Ω_0	longitude of ascending node at t_{oe}
i_0	inclination at t_{oe}
ω	argument of perigee
$\dot{\Omega}$	rate of Ω_0
$\dot{i}$	rate of i
C_{rs}, C_{rc}	correction coefficients for sine and cosine terms of r
C_{is}, C_{ic}	correction coefficients for sine and cosine terms of i
C_{us}, C_{uc}	correction coefficients for sine and cosine terms of ω

The truncated Z-count value in the HOW corresponds to the time of transmission of the next navigation data subframe. To get the time of transmission of the current subframe, the truncated Z-count should be multiplied by 6 and 6 s should be subtracted from the result.

Remaining Parameters   The remaining parameters of the navigation data are also decoded according to ICD-GPS-200 (1991), e.g., the ephemeris parameters are decoded according to Table 8.2. In the two's-complement, the sign bit ($+$ or $-$) occupies the MSB. The unit semicircle is multiplied with π to be converted into the unit radian.

The parameter IODE is short for Issue of Data Ephemerides. IODE is an eight-bit number that uniquely identifies the data set. All parameters are listed in Table 8.3 and explained in Section 8.3.

8.3   Computation of Satellite Position

This section connects Earth-centered and Earth-fixed (ECEF) coordinates X, Y, Z to a satellite position described in space by Keplerian orbit elements. First we recall the orbit elements a, e, ω, Ω, i, and μ listed in Figure 8.5. The six Keplerian orbit elements constitute an important description of the orbit, so they are repeated in schematic form in Table 8.4.

The rest of this section is unavoidably somewhat mathematical; many readers will proceed, assuming ECEF coordinates are found.

The X-axis points toward the intersection between equator and the Greenwich meridian. For our purpose this direction can be considered fixed. The Z-axis co-

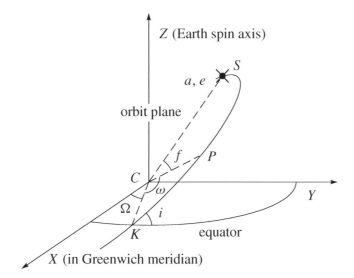

FIGURE 8.5. The Keplerian orbit elements: semi-major axis a, eccentricity e, inclination of orbit i, right ascension Ω of ascending node K, argument of perigee ω, and true anomaly f. Perigee is denoted P. The center of the Earth is denoted C.

incides with the spin axis of the Earth. The Y-axis is orthogonal to these two directions and forms a right-handed coordinate system.

The orbit plane intersects the Earth equator plane in the *nodal line*. The direction in which the satellite moves from south to north is called the *ascending node K*. The angle between the equator plane and the orbit plane is the *inclination i*. The angle at the Earth's center C between the X-axis and the ascending node K is called Ω; it is a right ascension. The angle at C between K and the perigee P is called *argument of perigee ω*; it increases counterclockwise viewed from the positive Z-axis.

Figure 8.6 shows a coordinate system in the orbital plane with origin at the Earth's center C. The ξ-axis points to the perigee and the η-axis toward the descending node. The ζ-axis is perpendicular to the orbit plane. From Figure 8.6 we

TABLE 8.4. Keplerian orbit elements: Satellite position

a	semi-major axis	size and shape of orbit
e	eccentricity	
ω	argument of perigee	the orbital plane in the apparent system
Ω	right ascension of ascending node	
i	inclination	
μ	mean anomaly	position in the plane

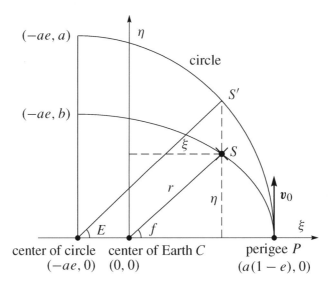

FIGURE 8.6. The elliptic orbit with (ξ, η) coordinates. The true anomaly f at C.

read the eccentric anomaly E and the true anomaly f. Also, immediately we have

$$\xi = r \cos f = a \cos E - ae = a(\cos E - e),$$
$$\eta = r \sin f = \tfrac{b}{a} a \sin E = b \sin E = a\sqrt{1 - e^2} \sin E.$$

Hence the position vector $\boldsymbol{r}$ of the satellite with respect to the center of the Earth C is

$$\boldsymbol{r} = \begin{bmatrix} \xi \\ \eta \\ \zeta \end{bmatrix} = \begin{bmatrix} a(\cos E - e) \\ a\sqrt{1 - e^2} \sin E \\ 0 \end{bmatrix}. \tag{8.1}$$

Simple trigonometry leads to the following expression for the norm:

$$\|\boldsymbol{r}\| = a(1 - e \cos E). \tag{8.2}$$

In general, E varies with time t while a and e are nearly constant. (There are long and short periodic perturbations to e, only short for a.) Recall that $\|\boldsymbol{r}\|$ is the geometric distance between satellite S and the Earth center $C = (0, 0)$.

For later reference we introduce the mean motion n, which is the mean angular satellite velocity. If the period of one revolution of the satellite is T, we have

$$n = \frac{T}{2\pi} = \sqrt{\frac{GM}{a^3}}. \tag{8.3}$$

The product GM was introduced on page 46 with the value $3.986\,005 \cdot 10^{14}\,\mathrm{m}^3/\mathrm{s}^2$. This value *shall* be used for computation of satellite positions (based on broadcast ephemerides), although more recent values of GM are available; see Section 8.10.

Let t_0 be the time the satellite passes perigee, so that $\mu(t) = n(t - t_0)$. Kepler's famous equation relates the mean anomaly μ and the eccentric anomaly E:

$$E = \mu + e \sin E. \tag{8.4}$$

From Equation (8.1) we finally get

$$f = \arctan \frac{\eta}{\xi} = \arctan \frac{\sqrt{1 - e^2} \sin E}{\cos E - e}. \tag{8.5}$$

By this we have connected the true anomaly f, the eccentric anomaly E, and the mean anomaly μ. These relations are basic for every calculation of a satellite position.

It is important to realize that the orbital plane remains fairly stable in relation to the geocentric X, Y, Z-system. In other words, seen from space, the orbital plane remains fairly fixed in relation to the equator. The Greenwich meridian plane rotates around the Earth spin axis in accordance with Greenwich apparent sidereal time (GAST), that is, with a speed of approximately 24 h/day. A GPS satellite performs two revolutions a day in its orbit having a speed of 3.87 km/s.

In the orbital plane the Cartesian coordinates of satellite S are given as

$$\begin{bmatrix} r_j^k \cos f_j^k \\ r_j^k \sin f_j^k \\ 0 \end{bmatrix},$$

where $r_j^k = \|\boldsymbol{r}(t_j)\|$ comes from (8.2) with a, e, and E evaluated for $t = t_j$; refer to Figure 8.5.

This vector is rotated into the X, Y, Z-coordinate system by the following sequence of 3D rotations:

$$R_3(-\Omega_j^k) R_1(-i_j^k) R_3(-\omega_j^k).$$

The matrix that rotates the XY-plane by φ, and leaves the Z-direction alone, is

$$R_3(\varphi) = \begin{bmatrix} \cos \varphi & \sin \varphi & 0 \\ -\sin \varphi & \cos \varphi & 0 \\ 0 & 0 & 1 \end{bmatrix} \tag{8.6}$$

and similarly for a rotation about the X-axis:

$$R_1(\varphi) = \begin{bmatrix} 1 & 0 & 0 \\ 0 & \cos \varphi & \sin \varphi \\ 0 & -\sin \varphi & \cos \varphi \end{bmatrix}. \tag{8.7}$$

Finally, the geocentric coordinates of satellite k at time t_j are given as

$$\begin{bmatrix} X^k(t_j) \\ Y^k(t_j) \\ Z^k(t_j) \end{bmatrix} = R_3(-\Omega_j^k) R_1(-i_j^k) R_3(-\omega_j^k) \begin{bmatrix} r_j^k \cos f_j^k \\ r_j^k \sin f_j^k \\ 0 \end{bmatrix}. \tag{8.8}$$

However, GPS satellites do not follow the presented normal orbit theory. We have to use time-dependent, more accurate orbit values. They come to us as the socalled broadcast ephemerides; see Section 8.2.2. We insert those values in a procedure given below and finally we get a set of variables to be inserted into (8.8).

Obviously, the vector is time-dependent, and one speaks about the *ephemeris* (plural: *ephemerides*, emphasis on "phem") of the satellite. These are the parameter values at a specific time. Each satellite transmits its unique ephemeris data.

The parameters chosen for description of the actual orbit of a GPS satellite and its perturbations are similar to the Keplerian orbital elements. The broadcast ephemerides are calculated using the immediate previous part of the orbit and they predict the following part of the orbit. The broadcast ephemerides are accurate to 1–2 m. For geodetic applications, better accuracy is needed. One possibility is to obtain post-processed *precise ephemerides*, which are accurate at the dm-level.

An ephemeris is intended for use from the epoch t_{oe} of reference counted in seconds of the GPS week. It is nominally at the center of the interval over which the ephemeris is useful. The broadcast ephemerides are intended for use during this period. However, they describe the orbit to within the specified accuracy for 2 hours afterward. The broadcast ephemerides include the parameters in Table 8.2. The coefficients C_ω, C_r, and C_i correct argument of perigee, orbit radius, and orbit inclination due to inevitable perturbations of the theoretical orbit caused by variations in the Earth's gravity field, albedo and sun pressure, and attraction from sun and moon.

Given the transmit time t (in GPS time), the following procedure gives the necessary variables to use in (8.8):

Time elapsed since t_{oe}
$$t_j = t - t_{\text{oe}}$$

Mean anomaly at time t_j
$$\mu_j = \mu_0 + \left(\sqrt{GM/a^3} + \Delta n \right) t_j$$

Iterative solution for E_j
$$E_j = \mu_j + e \sin E_j$$

True anomaly
$$f_j = \arctan \frac{\sqrt{1-e^2} \sin E_j}{\cos E_j - e}$$

Longitude for ascending node
$$\Omega_j = \Omega_0 + (\dot{\Omega} - \omega_e) t_j - \omega_e t_{\text{oe}}$$
$$\omega_e = 7.2921151467 \cdot 10^{-5} \, \text{rad/s}$$

Argument of perigee
$$\omega_j = \omega + f_j + C_{\omega c} \cos 2(\omega + f_j) + C_{\omega s} \sin 2(\omega + f_j)$$

Radial distance
$$r_j = a(1 - e \cos E_j) + C_{rc} \cos 2(\omega + f_j)$$
$$+ C_{rs} \sin 2(\omega + f_j)$$

Inclination
$$i_j = i_0 + \dot{i} t_j + C_{ic} \cos 2(\omega + f_j) + C_{is} \sin 2(\omega + f_j).$$

The mean Earth rotation is denoted ω_e. This algorithm is coded as the *M*-file satpos. The function calculates the position of any GPS satellite at any time. It is fundamental to every position calculation.

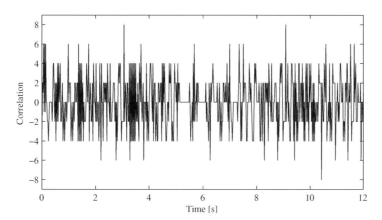

FIGURE 8.7. Correlation of 12 s data with the preamble.

To avoid under- or overflow at the beginning or end of a week, we use the *M*-file check_t.

8.4   Pseudorange Estimation

Pseudorange estimations can be divided into two sets of computations. The first computational method is to find the initial set of pseudoranges, and the second computational method is to keep track of the pseudoranges after the first set is estimated. Two computational methods are described below.

8.4.1   The Initial Set of Pseudoranges

To find the initial set of pseudoranges between the receiver and the tracked satellites, at least 12 s of data are needed. When the receiver has collected more than 12 s of data, the data are correlated with the preamble from the TLM word. The TLM word is located at the start of each subframe. Such a correlation is shown in Figure 8.7.

The length of the preamble is 8 samples, so if the TLM word is present in the data, there will be a correlation value of 8. It can be seen in Figure 8.7 that there is a correlation value corresponding to the preamble twice in the 12 s of data. This is used to check if it really is a subframe that is found or it were just a bit combination similar to the preamble. The correlation is carried out for all the tracked satellites.

After the preamble is identified, the start of a subframe is found for all the available satellites. In Figure 8.8, the start of the subframes is plotted for four channels.

It is known that the travel time from the satellites to the Earth is 65–83 ms. This is used to set the initial pseudorange. The satellite closest to the Earth is the satellite with the earliest arriving subframe. In this case the satellite in chan-

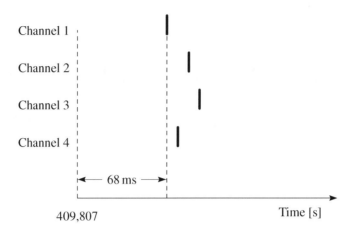

FIGURE 8.8. The transmit time and start of the subframe for four channels.

nel 1 has a travel time of 68 ms. The travel time of the remaining channels is then computed with respect to channel 1. In the example this leads to travel times and pseudoranges for the four channels as listed in Table 8.5.

In the present example, the time resolution is 1 ms, which corresponds to a pseudorange of 300,000 m. To make the pseudorange more useful, the tracking loop needs to find the start of the C/A code in the specific frame. This means that the time resolution is the sampling time, and in this case the sampling frequency is 38.192 MHz. This sampling frequency leads to a pseudorange accuracy of 8 m.

With the initial pseudoranges the receiver position can be computed. The output of that computation is a receiver position (X, Y, Z) and a receiver clock offset dt. The clock offset can be used to adjust the travel time of the reference satellite. In this case it was channel 1 with the travel time of 65 ms. And with this procedure the receiver can estimate the actual pseudoranges after two computations.

8.4.2   Estimation of Subsequent Pseudoranges

Section 8.4.1 describes how the initial pseudoranges are estimated. This subsection describes how the subsequent pseudoranges are computed.

TABLE 8.5. Initial pseudoranges for all tracked satellites

Channel	Travel time [ms]	Pseudorange [m]
1	65	19,486,509.8
2	74	22,184,641.9
3	81	24,283,189.1
4	69	20,685,679.6

When computing the subsequent pseudoranges, keep track of two issues. The first issue is the difference in ms between the start of the subframe compared to the reference satellite. The second issue is *the start of the C/A code, which gives the exact pseudorange for the channel.*

When the receiver is computing the subsequent pseudoranges, the receiver moves all the indexes 100 ms. (The receiver moves the indexes 100 ms if the receiver is set up to compute positions 10 times per s.) Then the start of the C/A code is found for all the new indexes for all the tracked channels. In this way it is possible to produce pseudoranges every millisecond and the receiver computes positions 1000 times per s.

8.5   Computation of Receiver Position

8.5.1   Time

The traveling time between satellite k and receiver i is denoted τ_i^k. Let c denote the velocity of light in vacuum, and the pseudorange P_i^k is defined as

$$t_i - t^k = \tau_i^k = P_i^k/c. \tag{8.9}$$

Let any epoch in GPS time (GPST) be called t^{GPS}. The clock at satellite k and the clock at receiver i do not run perfectly aligned with GPST. Thus, we introduce clock offsets defined as

$$t_i = t^{\text{GPS}} + dt_i, \tag{8.10}$$

$$t^k = (t_i - \tau_i^k)^{\text{GPS}} + dt^k. \tag{8.11}$$

In the latter equation we substitute $dt^k = a_0 + a_1(t^k - t_{\text{oe}}) + \cdots$ as given from the ephemeris:

$$(t_i - \tau_i^k)^{\text{gps}} = t^k - \big(a_0 + a_1(t^k - t_{\text{oe}}) + \cdots\big). \tag{8.12}$$

The left term is used as argument when computing the satellite position. Rearranging Equation (8.9), we get

$$t^k = t_i - P_i^k/c. \tag{8.13}$$

One way of using this equation is to consider t_i as given. That is the epoch time as defined in terms of the receiver clock. The pseudorange P_i^k is likewise known as an observable. Hence t^k can be computed, and after correcting for the satellite clock offset dt^k we obtain the transmit time in GPST. This is the procedure used for hardware receivers.

For software receivers the situation is a little different. The time t_{common} common to all pseudorange observations is defined as the time of transmission at the satellites. Hence the computation of position of satellite k is done at

$$t^k = t_{\text{common}} - dt^k. \tag{8.14}$$

The only "receiver time" used is the relative time of reception from each of the satellites and which makes the individual pseudorange.

A consequence of this time definition is that the computed satellite coordinates immediately refer to the ECEF system, and therefore satellite coordinates are not to be rotated about the Z-axis by an angle equal to the travel time times the Earth's rotation rate.

8.5.2   Linearization of the Observation Equation

The most commonly used algorithm for position computations from pseudoranges is based on the least-squares method. This method is used when there are more observations than unknowns. This section describes how the least-squares method is used to find the receiver position from pseudoranges to four or more satellites.

Let the geometrical range between satellite k and receiver i be denoted ρ_i^k, let c denote the speed of light, let dt_i be the receiver clock offset, let dt^k be the satellite clock offset, let T_i^k be the tropospheric delay, let I_i^k be the ionospheric delay, and let e_i^k be the observational error of the pseudorange. Then the basic observation equation for the pseudorange P_i^k is

$$P_i^k = \rho_i^k + c(dt_i - dt^k) + T_i^k + I_i^k + e_i^k. \tag{8.15}$$

The geometrical range ρ_i^k between the satellite and the receiver is computed as

$$\rho_i^k = \sqrt{(X^k - X_i)^2 + (Y^k - Y_i)^2 + (Z^k - Z_i)^2}. \tag{8.16}$$

Inserting (8.15) into (8.16) yields

$$P_i^k = \sqrt{(X^k - X_i)^2 + (Y^k - Y_i)^2 + (Z^k - Z_i)^2} + c(dt_i - dt^k) + T_i^k + I_i^k + e_i^k. \tag{8.17}$$

From the ephemerides—which include information on the satellite clock offset dt^k—the position of the satellite (X^k, Y^k, Z^k) can be computed. (The M-file sat-pos does the job.)

The tropospheric delay T_i^k is computed from an a priori model that is coded as tropo; the ionospheric delay I_i^k may be estimated from another a priori model, the coefficients of which are part of the broadcast ephemerides. The equation contains four unknowns X_i, Y_i, Z_i, and dt_i; the error term e_i^k is minimized by using the least-squares method. To compute the position of the receiver, at least four pseudoranges are needed.

Equation (8.17) is nonlinear with respect to the receiver position (X_i, Y_i, Z_i), so the equation has to be *linearized* before using the least-squares method.

We analyze the nonlinear term in (8.17)

$$f(X_i, Y_i, Z_i) = \sqrt{(X^k - X_i)^2 + (Y^k - Y_i)^2 + (Z^k - Z_i)^2}. \tag{8.18}$$

Linearization starts by finding an initial position for the receiver: $(X_{i,0}, Y_{i,0}, Z_{i,0})$. This is often chosen as the center of the Earth (0,0,0).

The increments ΔX, ΔY, ΔZ are defined as

$$X_{i,1} = X_{i,0} + \Delta X_i,$$
$$Y_{i,1} = Y_{i,0} + \Delta Y_i, \qquad (8.19)$$
$$Z_{i,1} = Z_{i,0} + \Delta Z_i.$$

They update the approximate receiver coordinates. So the Taylor expansion of $f(X_{i,0} + \Delta X_i, Y_{i,0} + \Delta Y_i, Z_{i,0} + \Delta Z_i)$ is

$$f(X_{i,1}, Y_{i,1}, Z_{i,1}) = f(X_{i,0}, Y_{i,0}, Z_{i,0}) + \frac{\partial f(X_{i,0}, Y_{i,0}, Z_{i,0})}{\partial X_{i,0}} \Delta X_i$$
$$+ \frac{\partial f(X_{i,0}, Y_{i,0}, Z_{i,0})}{\partial Y_{i,0}} \Delta Y_i + \frac{\partial (X_{i,0}, Y_{i,0}, Z_{i,0})}{\partial Z_{i,0}} \Delta Z_i. \qquad (8.20)$$

Equation (8.20) includes only first-order terms; hence the function determines an approximate position. The partial derivatives in Equation (8.20) are

$$\frac{\partial f(X_{i,0}, Y_{i,0}, Z_{i,0})}{\partial X_{i,0}} = -\frac{X^k - X_{i,0}}{\rho_i^k},$$
$$\frac{\partial f(X_{i,0}, Y_{i,0}, Z_{i,0})}{\partial Y_{i,0}} = -\frac{Y^k - Y_{i,0}}{\rho_i^k},$$
$$\frac{\partial f(X_{i,0}, Y_{i,0}, Z_{i,0})}{\partial Z_{i,0}} = -\frac{Z^k - Z_{i,0}}{\rho_i^k}.$$

Let $\rho_{i,0}^k$ be the range computed from the approximate receiver position; the first-order linearized observation equation becomes

$$P_i^k = \rho_{i,0}^k - \frac{X^k - X_{i,0}}{\rho_{i,0}^k} \Delta X_i - \frac{Y^k - Y_{i,0}}{\rho_{i,0}^k} \Delta Y_i - \frac{Z^k - Z_{i,0}}{\rho_{i,0}^k} \Delta Z_i$$
$$+ c(dt_i - dt^k) + T_i^k + I_i^k + e_i^k, \quad (8.21)$$

where we explicitly have

$$\rho_{i,0}^k = \sqrt{(X^k - X_{i,0})^2 + (Y^k - Y_{i,0})^2 + (Z^k - Z_{i,0})^2}. \qquad (8.22)$$

8.5.3   Using the Least-Squares Method

A least-squares problem is given as a system $A\boldsymbol{x} = \boldsymbol{b}$ with no solution. A has m rows and n columns, with $m > n$; there are more observations $b_1, \ldots, b_m$ than free parameters $x_1, \ldots, x_n$. The best choice, we will call it $\hat{\boldsymbol{x}}$, is the one that minimizes the length of the error vector $\hat{\boldsymbol{e}} = \boldsymbol{b} - A\hat{\boldsymbol{x}}$. If we measure this length in the usual way, so that $\|e\|^2 = (b - Ax)^{\mathrm{T}}(b - Ax)$ is the sum of squares of the m separate errors, minimizing this quadratic gives the normal equations

$$A^{\mathrm{T}}A\hat{x} = A^{\mathrm{T}}b \qquad \text{or} \qquad \hat{x} = (A^{\mathrm{T}}A)^{-1}A^{\mathrm{T}}b \tag{8.23}$$

and the error vector is

$$\hat{e} = b - A\hat{x}. \tag{8.24}$$

The covariance matrix for the parameters $\hat{x}$ is

$$\Sigma_{\hat{x}} = \hat{\sigma}_0^2 (A^{\mathrm{T}}A)^{-1} \qquad \text{with} \qquad \hat{\sigma}_0^2 = \frac{\hat{e}^{\mathrm{T}}\hat{e}}{m - n}. \tag{8.25}$$

The linearized observation equation (8.21) can be rewritten in a vector formulation

$$P_i^k = \rho_{i,0}^k + \begin{bmatrix} -\dfrac{X^k - X_{i,0}}{\rho_{i,0}^k} & -\dfrac{Y^k - Y_{i,0}}{\rho_{i,0}^k} & -\dfrac{Z^k - Z_{i,0}}{\rho_{i,0}^k} & 1 \end{bmatrix} \begin{bmatrix} \Delta X_i \\ \Delta Y_i \\ \Delta Z_i \\ c\,dt_i \end{bmatrix} - c\,dt^k + T_i^k + I_i^k + e_i^k. \tag{8.26}$$

We rearrange this to resemble the usual formulation of a least-squares problem $Ax = b$:

$$\begin{bmatrix} -\dfrac{X^k - X_{i,0}}{\rho_{i,0}^k} & -\dfrac{Y^k - Y_{i,0}}{\rho_{i,0}^k} & -\dfrac{Z^k - Z_{i,0}}{\rho_{i,0}^k} & 1 \end{bmatrix} \begin{bmatrix} \Delta X_i \\ \Delta Y_i \\ \Delta Z_i \\ c\,dt_i \end{bmatrix} = P_i^k - \rho_{i,0}^k + c\,dt^k - T_i^k - I_i^k - e_i^k. \tag{8.27}$$

A unique solution cannot be found from a single equation. Let $b_i^k = P_i^k - \rho_{i,0}^k + c\,dt^k - T_i^k - I_i^k - e_i^k$. Then the final solution comes from

$$Ax = \begin{bmatrix} -\dfrac{X^1 - X_{i,0}}{\rho_{i,0}^1} & -\dfrac{Y^1 - Y_{i,0}}{\rho_{i,0}^1} & -\dfrac{Z^1 - Z_{i,0}}{\rho_{i,0}^1} & 1 \\[2mm] -\dfrac{X^2 - X_{i,0}}{\rho_{i,0}^2} & -\dfrac{Y^2 - Y_{i,0}}{\rho_{i,0}^2} & -\dfrac{Z^2 - Z_{i,0}}{\rho_{i,0}^2} & 1 \\[2mm] -\dfrac{X^3 - X_{i,0}}{\rho_{i,0}^3} & -\dfrac{Y^3 - Y_{i,0}}{\rho_{i,0}^3} & -\dfrac{Z^3 - Z_{i,0}}{\rho_{i,0}^3} & 1 \\[2mm] \vdots & \vdots & \vdots & \vdots \\[2mm] -\dfrac{X^m - X_{i,0}}{\rho_{i,0}^k} & -\dfrac{Y^m - Y_{i,0}}{\rho_{i,0}^m} & -\dfrac{Z^m - Z_{i,0}}{\rho_{i,0}^m} & 1 \end{bmatrix} \begin{bmatrix} \Delta X_{i,1} \\ \Delta Y_{i,1} \\ \Delta Z_{i,1} \\ c\,dt_{i,1} \end{bmatrix} = b - e. \tag{8.28}$$

If $m \geq 4$, there is a unique solution: $\Delta X_{i,1}$, $\Delta Y_{i,1}$, $\Delta Z_{i,1}$. This has to be added to the approximate receiver position to get the next approximate position:

$$\begin{aligned} X_{i,1} &= X_{i,0} + \Delta X_{i,1}, \\ Y_{i,1} &= Y_{i,0} + \Delta Y_{i,1}, \\ Z_{i,1} &= Z_{i,0} + \Delta Z_{i,1}. \end{aligned} \tag{8.29}$$

TABLE 8.6. Typical standard deviations for pseudorange measurement

Error source	σ [m]
Satellite clock and orbit	1–2
Atmospheric modeling	4
Multipath and receiver noise	1

The next iteration restarts from (8.26) to (8.29) with $_{i,0}$ substituted by $_{i,1}$. These iterations continue until the solution $\Delta X_{i,1}$, $\Delta Y_{i,1}$, $\Delta Z_{i,1}$ is at meter level. Often two to three iterations are sufficient to obtain that goal; refer to Strang & Borre (1997).

If the observations (pseudoranges) and ephemerides are stored for later post-processing, a popular storage format is RINEX. A description of this can be found at http://www.ngs.noaa.gov/CORS/instructions2/.

When starting from RINEX format, the following M-files can be used for computation of the receiver position: easy3, get_eph, anheader, fepoch_0, fobs_typ, recpo_ls, find_eph, check_t, satpos, e_r_corr, topocent, tropo, and frgeod.

8.5.4 Real-Time Positioning Accuracy

A pseudorange observation on L_1 is typically influenced by several error sources: the broadcast orbits and the satellite clock offsets are not exact, the signal propagation through the atmosphere may only partly be modeled correctly, and the receiver adds some noise, and finally, signal multipath plays a role.

It is difficult to give precise estimates for the various errors; however, Table 8.6 indicates typical standard deviations for the said contributions.

8.6 Time Systems Relevant for GPS

The fundamental time interval unit is one SI second. The SI second was defined at the 13th general conference of the International Committee of Weights and Measures in 1967, as the "duration of 9,192,631,770 periods of the radiation corresponding to the transition between the two hyperfine levels of the ground state of the cesium 133 atom." The SI day is defined as 86,400 seconds and the Julian century as 36,525 days.

Since the apparent revolution of the sun about the Earth is nonuniform (this follows from Kepler's second law) a fictitious mean sun is defined that moves along the equator with uniform velocity. The hour angle of this fictitious sun is called Universal Time (UT).

The time epoch denoted by the Julian Date (JD) is expressed by a certain number of days and fraction of a day after a fundamental epoch sufficiently in the past

TABLE 8.7. Date for introduction of leap seconds to be added to UTC to get GPST

Total number of leap seconds	Date introduced
1	1 July 1981
2	1 July 1982
3	1 July 1983
4	1 July 1985
5	1 Jan. 1988
6	1 Jan. 1990
7	1 Jan. 1991
8	1 July 1992
9	1 July 1993
10	1 July 1994
11	1 Jan. 1996
12	1 July 1997
13	1 Jan. 1999
14	1 Jan. 2006

to precede the historical record, chosen to be at 12^h UT on January 1, 4713 BC. The Julian day number denotes a day in this continuous count, or the length of time that has elapsed at 12^h UT on the day designated since this epoch. The JD of the standard epoch of UT is called J2000.0, where

$$J2000.0 = JD\ 2,451,545.0 = 2000\ January\ 1.5^d\ UT.$$

The astronomic year commences at 0^h UT on December 31 of the previous year so that 2000 January 1.5^d UT = 2000 January 1 12^h. JD is a large number, so often it is replaced by the *Modified Julian Date* MJD:

$$MJD = JD - 2,400,000.5.$$

Hence J2000.0 = MJD 51,544.5. Note an MJD starts at midnight.

Because GPS Time (GPST) is a continuous-time scale, it does not maintain synchronization with the solar day since the Earth's rotation rate is slowing by an average of about 1 s per year. This problem is solved by defining Universal Time Coordinated (UTC), which runs at the same rate as GPST but is incremented by leap seconds periodically. Leap seconds are introduced by the IERS so that UTC does not vary from UT1 by more than 0.9 s. UT1 is UT corrected for polar motion. (IERS is an acronym for the International Earth Rotation Service. This service is also responsible for maintaining continuity with earlier data collected by optical instruments.) First preference is given to the end of June and December, and second preference to the end of March and September.

The time signals broadcast by the GPS satellites are synchronized with the atomic clock at the GPS Master Control Station in Colorado. GPST was set to 0^h

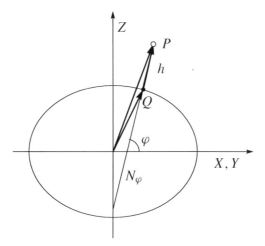

FIGURE 8.9. Conversion between (φ, λ, h) and Cartesian (X, Y, Z).

UTC on January 6, 1980, but is not incremented by UTC leap seconds. According to Table 8.7 there has been a total of 14 leap seconds since January 6, 1980, so that in early 2006

$$GPST = UTC + 14\,s.$$

GPS week numbers and seconds of week   Along with GPST, from the very beginning was introduced the GPS week numbers. Since January 6, 1980, any week has been designated its own number. At the time of writing we have week number 1315. To identify a given epoch within the week, the concept of seconds of week (SOW) is used. This number counts from midnight between Saturday and Sunday, which is also the beginning of the GPS week.

Furthermore, for convenience the individual days of the week are numbered: Sunday 0, Monday 1, Tuesday 2, Wednesday 3, Thursday 4, Friday 5, and Saturday 6.

Professional GPS softwares use the day of week for numerical reasons. SOW may be as large as $7 \times 24 \times 60 \times 60 = 604{,}800\,s$. In order to keep track of the mm in a point position we have to know time at the level of 0.01 nanosecond. Using seconds of week with 12 decimals is beyond the limits of most computers. So either you may split the real number holding the seconds of week into an integer part and a decimal part or you may compute time in terms of GPS week number, day of week, and seconds of day.

8.7   Coordinate Transformations

The immediate result of a satellite positioning is a set of X-, Y-, Z-values, which we most often want to convert to latitude φ, longitude λ, and height h, refer to Figure 8.9. We start by introducing the relation between Cartesian (X, Y, Z) co-

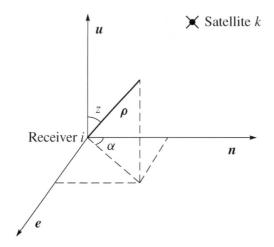

FIGURE 8.10. Zenith distance z and azimuth α in the topocentric system (e, n, u).

ordinates and geographical (φ, λ, h) coordinates:

$$X = (N_\varphi + h) \cos \varphi \cos \lambda, \tag{8.30}$$

$$Y = (N_\varphi + h) \cos \varphi \sin \lambda, \tag{8.31}$$

$$Z = \left((1 - f)^2 N_\varphi + h\right) \sin \varphi. \tag{8.32}$$

The *reference ellipsoid* is the surface given by $X^2 + Y^2 + \left(\frac{a}{b}\right)^2 Z^2 = a^2$. The radius of curvature in the prime vertical (which is the vertical plane normal to the astronomical meridian) is given as

$$N_\varphi = \frac{a}{\sqrt{1 - f(2 - f) \sin^2 \varphi}}. \tag{8.33}$$

In spherical approximation, i.e., $f = 0$, the unit vector u normal to the surface is

$$u = \begin{bmatrix} \cos \varphi \cos \lambda \\ \cos \varphi \sin \lambda \\ \sin \varphi \end{bmatrix}. \tag{8.34}$$

The tangent n and binormal e unit vectors are derivatives of u (n alludes to northing, e to easting, and u to up, i.e., the normal direction to the surface):

$$n = \frac{\partial u}{\partial \varphi} = \begin{bmatrix} -\sin \varphi \cos \lambda \\ -\sin \varphi \sin \lambda \\ \cos \varphi \end{bmatrix} \quad \text{and} \quad e = \frac{1}{\cos \varphi} \frac{\partial u}{\partial \lambda} = \begin{bmatrix} -\sin \lambda \\ \cos \lambda \\ 0 \end{bmatrix}. \tag{8.35}$$

Verify that $e = n \times u$.

The unit vectors n, e, u provide the natural coordinate frame at a point on the reference ellipsoid; see Figure 8.10. For reasons of reference we collect the three unit vectors e, n, u into the orthogonal matrix

$$F = \begin{bmatrix} e & n & u \end{bmatrix} = \begin{bmatrix} -\sin\lambda & -\sin\varphi\cos\lambda & \cos\varphi\cos\lambda \\ \cos\lambda & -\sin\varphi\sin\lambda & \cos\varphi\sin\lambda \\ 0 & \cos\varphi & \sin\varphi \end{bmatrix}. \qquad (8.36)$$

Example 8.1 Let P be given by the coordinates (φ, λ, h) in the WGS 84 system:

$$\varphi = \quad 40°\,07'\,04.595\,51'',$$

$$\lambda = 277°\,01'\,10.221\,76'',$$

$$h = 231.562\,\text{m}.$$

The longitude λ runs eastward, from $0°$ to $360°$.

We seek the (X, Y, Z) coordinates of P. The result is achieved by the M-file g2c:

$$X = \quad 596{,}915.961\,\text{m},$$

$$Y = -4{,}847{,}845.536\,\text{m},$$

$$Z = \quad 4{,}088{,}158.163\,\text{m}.$$

The reverse problem—compute (φ, λ, h) from (X, Y, Z)—requires an iteration for φ and h. Directly $\lambda = \arctan(Y/X)$. There is quick convergence for $h \ll N_\varphi$, starting at $h = 0$:

$$\varphi \text{ from } h \text{ (8.32):} \qquad \varphi = \arctan\left(\frac{Z}{\sqrt{X^2 + Y^2}}\left(1 - \frac{(2-f)fN_\varphi}{N_\varphi + h}\right)^{-1}\right),$$
$$\qquad (8.37)$$

$$h \text{ from } \varphi \text{ (8.30)–(8.31):} \qquad h = \frac{\sqrt{X^2 + Y^2}}{\cos\varphi} - N_\varphi. \qquad (8.38)$$

For large h (or φ close to $\pi/2$) we recommend the procedure given in the M-file c2gm.

8.8   Universal Transverse Mercator Mapping

The geographical coordinates (φ, λ) locate a point on the reference ellipsoid. For many practical purposes it is useful to have a coordinate representation in the two-dimensional plane. The mapping of an ellipsoid into a plane may be done by a *conformal mapping*. Conformity leaves the shape of small figures while distances must be scaled.

The most widely spread conformal mapping was introduced by the Department of Defense in United States shortly after the Second World War. It is called the

Universal Transverse Mercator Grid System (UTM). The reference ellipsoid is the International Ellipsoid of 1924. The mapping is defined for the whole Earth. It was a desire to limit the variation of the scale m so the total mapping of the Earth is divided into 60 sections that are named zones. Each zone covers $6°$ in longitude, and they are numbered from 1 to 60. Number 1 covers $180°$–$174°$ West, number 2 covers $174°$–$168°$ West, and so on. The scale at the central meridian is $m_0 = 0.9996$. The UTM system is limited by the parallels $84°$ North and $80°$ South. The polar regions are mapped stereographically.

The coordinates are called *northing N* and *easting E*. Each zone has its own coordinate system. The central meridian has a false easting (FE) of 500,000 m and the equator has a false northing (FN) of 0 m for points on the northern hemisphere and a false northing of 10,000,000 m for points on the southern hemisphere. This simple arrangement leaves all coordinates positive.

The transformation of geographical coordinates (φ, λ) into UTM coordinates (N, E) and reversely appears often in practice. We point to the M-files geo2utm and utm2geo:

> [N, E] = geo2utm(phi, lambda, zone)
> [phi, lambda] = utm2geo(N, E, zone)

8.9   Dilution of Precision

The covariance matrix $\Sigma_{\hat{x}}$ as described in (8.25) contains information about the geometric quality of the position determination. It is smaller (and $\hat{x}$ is more accurate) when the satellites are well spaced.

The covariance matrix $\Sigma_{\hat{x}}$ is a 3 by 3 matrix in case we estimate (X, Y, Z) and a 4 by 4 matrix in case we estimate $(X, Y, Z, c\,dt)$. It is positive definite, so its inverse exists and is likewise positive definite. We introduce a local coordinate system with origin at $(\hat{X}, \hat{Y}, \hat{Z})$ and with axes parallel to the original ones. In the local system, let the point $x = (x, y, z)^{\mathrm{T}}$ lie on a surface described by the quadratic form

$$x^{\mathrm{T}} \Sigma_{\hat{x}}^{-1} x = c^2 \tag{8.39}$$

or equivalently, if $\Sigma_{\hat{x}}^{-1}$ is diagonal,

$$\frac{x^2}{c^2 \sigma_x^2} + \frac{y^2}{c^2 \sigma_y^2} + \frac{z^2}{c^2 \sigma_z^2} = 1. \tag{8.40}$$

If $\Sigma_{\hat{x}}^{-1}$ is nondiagonal, it can be brought on diagonal form in a rotated coordinate system.

The surface is an ellipsoid because $\Sigma_{\hat{x}}^{-1}$ is positive definite. It is the *confidence ellipsoid* of the point. Let $1 - \alpha$ be the probability that the correct position falls within the ellipsoid. With $1-\alpha = 0.95$, c^2 is given as $\chi^2_{3,1-\alpha} = 7.81$. In MATLAB this number is computed as chi2inv(0.95,3). The magnification factor c of the confidence ellipsoid is $c = \sqrt{7.81} = 2.80$.

If we want to improve the likelihood that a new sample lies within the ellipsoid, we may increase the probability $1 - \alpha$, the axes of the confidence ellipsoid, the c-value, or all of them.

We start from the covariance matrix of the least-squares problem (8.28):

$$\Sigma_{\text{ECEF}} = \left[\begin{array}{ccc|c} \sigma_X^2 & \sigma_{XY} & \sigma_{XZ} & \sigma_{X,c\,dt} \\ \sigma_{YX} & \sigma_Y^2 & \sigma_{YZ} & \sigma_{Y,c\,dt} \\ \sigma_{ZX} & \sigma_{ZY} & \sigma_Z^2 & \sigma_{Z,c\,dt} \\ \hline \sigma_{c\,dt,X} & \sigma_{c\,dt,Y} & \sigma_{c\,dt,Z} & \sigma_{c\,dt}^2 \end{array} \right]. \tag{8.41}$$

The law of covariance propagation transforms Σ_{ECEF} into the covariance matrix expressed in a local system with coordinates (e, n, u). The interesting 3 by 3 submatrix S of Σ_{ECEF} is shown in (8.41). After the transformation with F, the submatrix becomes

$$\Sigma_{\text{enu}} = \begin{bmatrix} \sigma_e^2 & \sigma_{en} & \sigma_{eu} \\ \sigma_{ne} & \sigma_n^2 & \sigma_{nu} \\ \sigma_{ue} & \sigma_{un} & \sigma_u^2 \end{bmatrix} = F^{\mathsf{T}} S F. \tag{8.42}$$

In practice we meet several forms of the *dilution of precision* (abbreviated DOP):

Geometric: $\displaystyle \text{GDOP} = \sqrt{\frac{\sigma_e^2 + \sigma_n^2 + \sigma_u^2 + \sigma_{c\,dt}^2}{\sigma_0^2}} = \sqrt{\frac{\text{tr}(\Sigma_{\text{ECEF}})}{\sigma_0^2}},$

Horizontal: $\displaystyle \text{HDOP} = \sqrt{\frac{\sigma_e^2 + \sigma_n^2}{\sigma_0^2}},$

Position: $\displaystyle \text{PDOP} = \sqrt{\frac{\sigma_e^2 + \sigma_n^2 + \sigma_u^2}{\sigma_0^2}} = \sqrt{\frac{\sigma_X^2 + \sigma_Y^2 + \sigma_Z^2}{\sigma_0^2}} = \sqrt{\frac{\text{tr}(\Sigma_{\text{enu}})}{\sigma_0^2}},$

Time: $\text{TDOP} = \sigma_{c\,dt}/\sigma_0,$

Vertical: $\text{VDOP} = \sigma_U/\sigma_0.$

Note that all DOP values are dimensionless. They multiply the range errors to give the position errors (approximately). Furthermore, we have

$$\text{GDOP}^2 = \text{PDOP}^2 + \text{TDOP}^2 = \text{HDOP}^2 + \text{VDOP}^2 + \text{TDOP}^2.$$

Some satellite constellations are better than others and the knowledge of the time of best satellite coverage is a useful tool for anybody using GPS. Experience shows that *good observations are achieved when PDOP < 5 and measurements come from at least five satellites.*

Example 8.2 To emphasize the fundamental impact of the matrix F in (8.36) we shall determine the elevation angle for a satellite. The local *topocentric system* uses three unit vectors $(e, n, u) = $ (east, north, up). Those are the columns of F. The vector r between satellite k and receiver i is

$$r = \left(X^k - X_i, Y^k - Y_i, Z^k - Z_i \right).$$

The unit vector in this satellite direction is $\rho = r/\|r\|$. Then Figure 8.10 gives

$$\rho^T e = \sin \alpha \sin z,$$
$$\rho^T n = \cos \alpha \sin z,$$
$$\rho^T u = \cos z.$$

From this we determine α and z. Especially we have $\sin h = \cos z = \rho^T u$ for the elevation angle h. The above formulas are the basis for the M-file topocent.

The angle h or rather $\sin h$ is an important parameter for any procedure calculating the tropospheric delay, cf. the M-file tropo.

Furthermore, the quantity $\sin h$ has a decisive role in planning observations: when is h larger than $15°$, say? Those are the satellites we prefer to use in GPS. Many other computations and investigations involve this elevation angle h.

Often we want to transform topocentric coordinate differences (x, y, z) to local coordinates (e, n, u). This transformation is achieved by F^T:

$$\begin{bmatrix} e \\ n \\ u \end{bmatrix} = F^T \begin{bmatrix} x \\ y \\ z \end{bmatrix}. \tag{8.43}$$

This transformation is implemented in the M-file cart2utm. Correspondingly, the covariance matrix Σ_{enu} is given as

$$\Sigma_{\text{enu}} = F^T \Sigma_{\hat{x}} F. \tag{8.44}$$

Example 8.3 A receiver position is $(3,435,470.80,\ 607,792.32,\ 5,321,592.38)$ with the following covariance matrix, unit m^2:

$$\Sigma_{\hat{x}} = \begin{bmatrix} 25 & -7.970 & 18.220 \\ -7.970 & 4 & -6.360 \\ 18.220 & -6.360 & 16 \end{bmatrix}.$$

We know the point has $\varphi = 56° \ 55'$, $\lambda = 10° \ 02'$, and we get

$$F^T = \begin{bmatrix} -0.1742 & 0.9847 & 0 \\ -0.8252 & -0.1460 & 0.5457 \\ 0.5374 & 0.0951 & 0.8380 \end{bmatrix};$$

hence the covariance matrix for the local coordinates is according to (8.44):

$$\Sigma_{\text{enu}} = \begin{bmatrix} 7.37 & 4.14 & -13.96 \\ 4.14 & 4.56 & -9.38 \\ -13.96 & -9.38 & 33.07 \end{bmatrix}.$$

From this we get $\sigma_e = 2.7$ m, $\sigma_n = 2.1$ m, and $\sigma_u = 5.7$ m.

8.10   World Geodetic System 1984

The ellipsoid in WGS 84 is defined through four parameters; see Anonymous (1997):

1. the semi-major axis $a = 6{,}378{,}137$ m,

2. the Earth's gravitational constant (including the mass of the Earth's atmosphere) $GM = 3{,}986{,}004.418 \times 10^8$ m^3/s^2,

3. the flattening $f = 1/298.257223563$,

4. the Earth's rotational rate $\omega = 7{,}292{,}115 \times 10^{-11}$ rad/s.

The International Astronomical Union uses $\omega_e = 7{,}292{,}115.1467 \times 10^{-11}$ rad/s, with four extra digits, together with a new definition of time, and this value for ω_e is used for GPS. The speed of light in vacuum is taken as

$$c = 299{,}792{,}458 \text{ m/s}.$$

Conceptually, WGS 84 is a very special datum as it includes a model for the gravity field. The description is given by spherical harmonics up to degree and order 180. This adds 32,755 more coefficients to WGS 84 allowing for determination of the global features of the geoid. A truncated model ($n = m = 18$) of the geoid is shown in Figure 8.11. For a more detailed description; see Anonymous (1997).

In North America the transformation from NAD 27 to WGS 84 is given as

$$\begin{bmatrix} X_{\text{WGS 84}} \\ Y_{\text{WGS 84}} \\ Z_{\text{WGS 84}} \end{bmatrix} = \begin{bmatrix} X_{\text{NAD 27}} + 9 \text{ m} \\ Y_{\text{NAD 27}} - 161 \text{ m} \\ Z_{\text{NAD 27}} - 179 \text{ m} \end{bmatrix}.$$

A typical datum transformation into WGS 84 only includes changes in the semi-major axis of the ellipsoid and its flattening and three translations of the origin of the ellipsoid.

WGS 84 is a global datum, allowing us to transform between regions by means of GPS. The importance of WGS 84 is undoubtedly to provide a *unified global datum*.

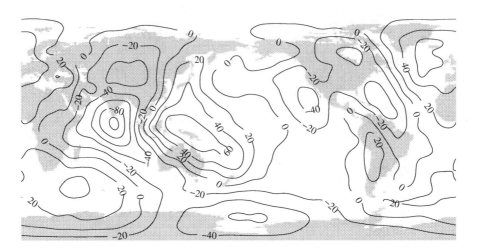

FIGURE 8.11. The WGS 84 geoid computed for a spherical harmonics expansion including terms with degree and order 18. The contour interval is 20 m.

8.11 Time and Coordinate Reference Frames for GPS and Galileo

GPS uses a *time reference frame* called GPS Time (GPST), which is steered by a real-time representation of Coordinated Universal Time (UTC) produced by the U.S. Naval Observatory. For Galileo an independent time scale, Galileo System Time (GST) will be established. The offset Δt between GPST and GST will probably be in the order of tens of ns.

A combined GPS and Galileo receiver will correct Galileo pseudoranges to GST and GPS pseudoranges to GPST. The offset Δt between GST and GPST can be estimated as an additional unknown in the position solution. Or equivalently we estimate the receiver clock offset dt to GPST and the receiver clock offset δt to GST. Now $\Delta t = dt - \delta t$.

The linearized system of observation equations for three GPS pseudoranges and two Galileo pseudoranges is as follows (the proposed RINEX version 3 uses the letter G for GPS and the letter E for Galileo):

$$
A\boldsymbol{x} =
\begin{bmatrix}
-\dfrac{X^{G1}-X_{i,0}}{\rho_i^{G1}} & -\dfrac{Y^{G1}-Y_{i,0}}{\rho_i^{G1}} & -\dfrac{Z^{G1}-Z_{i,0}}{\rho_i^{G1}} & 1 & 0 \\[2ex]
-\dfrac{X^{G2}-X_{i,0}}{\rho_i^{G2}} & -\dfrac{Y^{G2}-Y_{i,0}}{\rho_i^{G2}} & -\dfrac{Z^{G2}-Z_{i,0}}{\rho_i^{G2}} & 1 & 0 \\[2ex]
-\dfrac{X^{G3}-X_{i,0}}{\rho_i^{G3}} & -\dfrac{Y^{G3}-Y_{i,0}}{\rho_i^{G3}} & -\dfrac{Z^{G3}-Z_{i,0}}{\rho_i^{G3}} & 1 & 0 \\[2ex]
-\dfrac{X^{E1}-X_{i,0}}{\rho_i^{E1}} & -\dfrac{Y^{E1}-Y_{i,0}}{\rho_i^{E1}} & -\dfrac{Z^{E1}-Z_{i,0}}{\rho_i^{E1}} & 0 & 1 \\[2ex]
-\dfrac{X^{E2}-X_{i,0}}{\rho_i^{E2}} & -\dfrac{Y^{E2}-Y_{i,0}}{\rho_i^{E2}} & -\dfrac{Z^{E2}-Z_{i,0}}{\rho_i^{E2}} & 0 & 1
\end{bmatrix}
\begin{bmatrix}
\Delta X_{i,1} \\
\Delta Y_{i,1} \\
\Delta Z_{i,1} \\
c\,dt_{i,1} \\
c\,\delta t_{i,1}
\end{bmatrix}
= \boldsymbol{b} - \boldsymbol{e}.
$$

$$(8.45)$$

The *coordinate reference frame* for GPS is WGS 84. In fact, it is a realization of an International Terrestrial Reference Frame. It is realized through Cartesian coordinates and velocities of a global set of sites; these are identical to the tracking stations for GPS. Occasionally, the tracking stations are recoordinated. The latest modification was implemented in GPS week number 1150.

In a similar way, Galileo will use another independent realization (GTRF). In practice, both frames will be the same, but the difference is still a few centimeters, which may be significant for very precise applications.

Problems

1. Let the Fourier transform of $f(t)$ be given by

$$F(\omega) = \frac{1}{\alpha + j\omega} = \frac{\alpha}{\alpha^2 + \omega^2} - j\frac{\omega}{\alpha^2 + \omega^2} = \frac{1}{\sqrt{\alpha^2 + \omega^2}} \exp\left(-j \arctan\left(\frac{\omega}{\alpha}\right)\right).$$

Plot the Fourier transform (for different values of α) by

(a) the real and imaginary part of the Fourier transform,

(b) the amplitude and phase of the Fourier transform.

2. A pseudonoise sequence is generated using a feedback shift register of length $m = 10$. The chip rate is 1.023 Mchip/s. Find the following parameters:

(a) the pseudonoise sequence length,

(b) the chip duration of the pseudonoise sequence, and

(c) the pseudonoise sequence period.

3. The GPS signal transmitted from satellite k is defined by (2.3).

(a) Show how under ideal conditions we can obtain the in-phase I and quadrature Q signals from (2.3).

(b) Show how under ideal conditions we can obtain the navigation data $D^k(t)$ from the in-phase I and quadrature Q signals obtained in (a).

4. Within Chapter 4, the thermal noise power was computed for a 2 MHz bandwidth (approximate null-null bandwidth of the GPS C/A code signal) at 290° K (typically considered to be ambient temperature).

(a) What would the noise power be for a temperature of 100°K?

(b) What would the noise power be if the two primary lobes of the Galileo L1 BOC (1,1) signal were utilized to define the bandwidth while maintaining the 290°K temperature?

Comment on the dependence of the resulting noise power from changes in temperature and bandwidth.

5. What is the noise figure of the system depicted in Figure 4.2? What would it be if a passive antenna were utilized (move the filter and amplifier within the antenna to after the RF cable)?

6. Assuming the received GPS signal power from each satellite is −160 dBW, how many satellites must be a collected data set where the front end has a 2 dB noise figure and the C/A code sinc spectrum appears 2 dB above the filter shape in the collected data? Now redo the computation assuming the same 2 dB above the filter shape but now all the power results from a single satellite?

7. Assume it is possible to implement an ideal bandpass filter. Further assume that the transmitted signal has infinite bandwidth.

(a) What filter bandwidth is required to capture 98% of the GPS C/A code power? 98% of the Galileo L1 BOC(1,1) power?

(b) If the filter was designed to capture the first main spectrum lobe [lobes for the Galileo BOC(1,1) signal] what percent of the total power would that provide for the GPS C/A code signal? the Galileo L1 BOC(1,1)?

8. The collected data from the front end are represented as 8-bit samples (signed char format). Develop an M-file to convert this to

(a) 1 bit (± 1) values,

(b) 2 bit ($\pm 1, \pm 3$).

9. The M-file codegen generates any of the 32 PRN codes used in GPS. Compute the autocorrelation function for PRN 1 and notice the maximum and minimum values. Plot the resulting correlation function.

Hint: The correlation function between two sequences can be computed as

$$R_{xy}(n) = \sum_{l=0}^{1022} x(l)\, y(l+n),$$

or you may use the M-function xcorr.

10.   Compute the cross-correlation function between PRN1 and PRN2, and notice the maximum and minimum values.

Plot the resulting correlation function.

11.   Load the *M*-file unknown_prn.mat, and examine which of the 32 GPS satellites transmitted this code.

Hint: Use correlation! Determine the code phase (code phase is the location of the beginning of the PRN code in a sequence) of the unknown code.

12.   Create an *M*-function that generates a GPS C/A code. The function should have the PRN as input and should return the 1023-long sequence. The code generator should be based on the block diagram in Figure 2.5. Compare the results to the codes in Problems 9 and 10.

Hint: Use the following procedure:

– Select the G2 code phase selectors S1 and S2 based on the input PRN and the code phase selector column in Table 2.3.

– Initialize G1 and G2 registers (set all values to 1).

– Make a loop that runs until the registers have changed state to their initial value (all ones):

– generate XOR output from G1;

– generate XOR output from G2;

– create the output value of the generator;

– shift the values of the registers;

– change all values to −1's and 1's.

13.   Write an *M*-script that generates sine and cosine carriers. Generate two 1 ms sequences with

– sampling frequency $f_s = 10$ MHz,

– carrier frequency $f_c = 1.2$ MHz.

Plot the first 50 samples of the sequences and notice the discrete-time signals.

14.   Make a serial search algorithm based on the block diagram in Figure 6.1. Use the *M*-file data.mat, which is real GPS data sampled at a frequency of 38.192 MHz. The signal is downconverted to a nominal carrier frequency of 9.548 MHz. Use the sampled C/A codes for satellites 1, 3, and 6 from the *M*-files gold1.mat, gold3.mat, and gold6.mat. These *M*-files each contain a 1023 × 38,192 array corresponding to 1023 different versions of the C/A code with different code phases each of them with a length of 38,192 corresponding to the length of a complete code (1 ms with a sampling frequency of 38.192 MHz). Let the algorithm search for satellites 1, 3, and 6. Plot the result using the surf function.

Hint: Use the fastest computer available because the computations are heavy. Start MATLAB with matlab −nojvm −nosplash to increase performance. Load the data with

```
load data.mat;
x = double(data');
```

Load the C/A sequences with

```
load gold1.mat;
code = double(code);
```

15. Run the same scenario with the parallel code phase search algorithm found in the *M*-file acquisitionConvolution.m. Compare the results and performance of the two algorithms.

16. *Tracking only the C/A-code in the signal.* Load the *M*-file data_long.mat. This *M*-file contains the same data sequence as data.mat, except it is just about 30 s long. Use the block diagram in Figure 7.12 to track the code in the incoming signal data_long.mat.

Keep the frequency of the "Local Oscillator" constant at the frequency at which you have acquired the signal. This means that the phase of the local carrier is *not* locked, which implies that if the frequency is not 100% correct, the signal power will be switching between the In-phase and Quadrature arm.

The "PRN Code Generator" from Problem 12 should be used to produce three local code replicas. There should be 1 chip between the early and prompt replica. One chip corresponds to about 38 samples at 38.192 MHz.

- Make an acquisition on the first 1 ms of data_long.mat.

- Track the first 1 ms by using the scheme above, and use the "frequency" and "code phase" from the acquisition.

- After the first ms, construct the Code Lock Loop discriminator from the six outputs as
$$D = I_P(I_E - I_L) + Q_P(Q_E - Q_L).$$

- Use the output of the discriminator to adjust the "code phase" in the PRN Code Generator and continue to track the signal in 1-ms blocks.

Hints: When the code tracking is working, the sum of the squared prompt outputs is higher than the sum of the early and late outputs. This means that the local prompt code replica has a larger correlation value than the other two replicas.

17. Load the *M*-file navigation.mat. It contains synchronized navigation data bits represented by $[-1 \quad 1]$. Find the first beginning of a subframe in the navigation data sequence.

Hint: Use correlation with the preamble $[1 \ -1 \ -1 \ -1 \quad 1 \ -1 \quad 1 \quad 1]$. Notice that the sequence can be inverted!

18. Find the bits representing the Z-count and determine the time of week in days, hours, minutes, and seconds.

Hint: Look in the GPS Interface Control Document (icd200c.pdf) on page 65+ for a complete description of navigation data contents. Check if the sequence is inverted compared to the result of Problem 17.

19. We load the data file data_long2.mat. This M-file contains data collected from a local antenna and the sampling frequency is 38.192 MHz. Compute a position based on the data in the M-file.

Hints:

- make acquisition on the first ms of data,
- track all the available satellites,
- decode the navigation messages on all the satellites,
- compute pseudoranges to all the available satellites,
- compute the satellite positions at transmit time,
- compute the position based on the pseudoranges and the satellite positions.

20. Given a point with $(\varphi, \lambda, h) = (40°N, 277°E, 231\,m)$ in WGS 84, find the Cartesian coordinates (X, Y, Z). Hint: Use geo2cart.

21. Given a point with the (X, Y, Z) from Problem 20, find the geographical coordinates of the point in WGS 84. Hint: Use cart2geo.

22. Given $X = 3,429,122.662$, $Y = 604,646.845$, and $Z = 5,325,950.420$, convert (X, Y, Z) into (N, E) in UTM, zone 32. Hint: Use cart2utm.

23. Convert $N = 6,318,036.28$, $E = 560,828.13$ into geographical coordinates (φ, λ). Hint: Use utm2geo.

24. Convert $\varphi = 57°N$, $\lambda = 10°E$ into (N, E) in UTM, zone 32. Hint: Use geo2utm.

25. Convert the UTM coordinates $N = 6,317,972.081$, $E = 560,749.622$ into geographical coordinates. Hint: Use utm2geo.

26. Compute GDOP for a constellation of four satellites:

- one at zenith and three equally spread along the horizon;
- consider a constellation of four satellites equally spread at an elevation angle of 45°.

27. Convert the epoch 8h on 29 March 2005 to Modified Julian Date and GPS weeks and seconds of week. Hint: Use julday and gps_time.

MATLAB Code

A.1 Structure of the Code

The generic one-channel receiver is shown in Figure 5.1. The actual data flow and the MATLAB functions used by the software receiver are depicted in Figure A.1.

Below we mention a short description of the single structures of the receiver and the variables that conduct the behavior of the software receiver.

First, a small section (few ms) at the start of the data file is read and is passed to the acquisition M-file. The acquisition M-file looks for any presence of GPS signals. It estimates the frequency and the C/A code phase for every present GPS signal. The results are stored in the structure acqResults.

Next, the function preRun reads the acquisition results and initializes all software channels. If the number of available satellites is less than the number of channels, the unused channels are disabled. The same function also clears all processing results from possible previous runs. Thus, it prepares a clean environment for the following run.

After initialization of channels, a block of signal samples is read from the recorded file and is passed to the tracking function track. The tracking function tracks signals for all enabled channels, detects bit boundaries, stores bits of the navigation data, and decodes the data. Decoded ephemerides are stored in the structure eph. Tracking results (outputs from correlators, discriminators, etc.) are stored in the structure trackResults. In settings one may specify for how long the tracking should proceed.

After the tracking is finished, the function postNavigation is launched to process the signal. It identifies the start of a subframe, determines the signal trans-

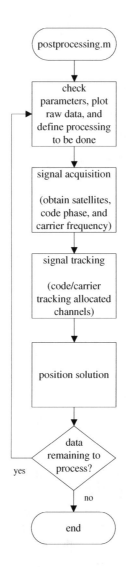

FIGURE A.1. GNSS software receiver flow diagram.

mission time, and estimates all pseudoranges. Then the M-file computes ECEF coordinates of the software antenna and the ECEF coordinates are converted to a specified coordinate system, say UTM or WGS84.

Finally, the results of acquisition, tracking, and positioning are plotted.

A.2   The settings Structure

All variables common to all receiver blocks, and block-specific variables are stored in one structure named settings. This approach enables a centralized and flexible management of the software. For instance, sampling frequency-based parameters are used by many files—from acquisition to pseudorange computation. Once the parameters are updated in the variable settings, all files will use the changed values.

One more advantage is that the function parameter list does not depend on how many parameters the function actually is using. Changes inside the function will not affect the calling function code.

The most commonly used variables are

IFfrequency Intermediate frequency of the GPS signal, Hz

samplingFrequency Frequency at which the GPS signal is sampled, Hz

msToProcess This variable is set to 37,000 to ensure all five 6-s subframes are processed and included in the output, the first 1000 ms can be excluded for any transient response

processBlockSize Size of the block to be processed by the tracking function

numberOfChannels Sets the number of channels of the software receiver.

The function initSettings creates the structure settings. The very first time this function is executed by the script init. The function should be executed every time variables are changed. An experienced user might change some of the variables directly in the settings structure. However, care must be taken as some variables have internal dependencies and are recomputed automatically. So it is safest to change variables in the function initSettings, which must be re-executed afterward.

The block-dependent variables are described in the following sections.

A.3   Acquisition Function

The function acquisition employs the parallel code phase search acquisition algorithm described in Section 6.4. The purpose is to find signal parameters for all available satellites in a few-ms-long data record. The implementation is based on the block diagram shown in Figure 6.8. The flow diagram for the actual code is shown in Figure A.2.

The acquisition function looks for a GPS signal in frequency steps of 0.5 kHz. For each frequency step, a parallel code search is performed. The correlation results are saved and the function proceeds with the next frequency step. Thus, the function steps through all frequency bands (user-defined Doppler space). Next the function looks for a maximum correlation value (correlation peak) in results from all frequency bins. After the peak is detected, the function looks for the second-highest correlation peak in the same frequency bin of the highest peak. Then the ratio of the two peaks is used for the signal detection rule. The ratio is compared to the value preset in the receiver variable acq_threshold.

The detector does not depend on sampling frequency and therefore is not dependent on the size of peak and noise level.

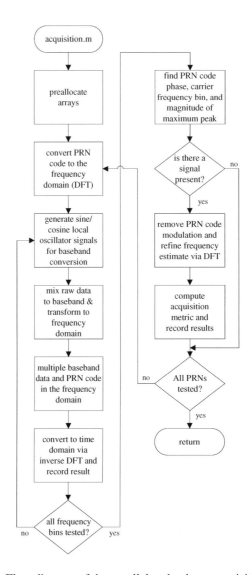

FIGURE A.2. Flow diagram of the parallel code phase acquisition algorithm.

If the value of the peak ratio is larger than a specified value, the fine carrier frequency is found via a postcorrelation FFT approach. This must be done to help the PLL in the tracking loop to start tracking the signal. A frequency accuracy of 0.5 kHz is too coarse for the PPL to start tracking.

The function parameters are an initial data record, a table with pregenerated C/A codes, and the settings structure.

The list of acquisition-specific variables contained in the structure settings is

acq_satelliteList A set of satellite PRNs can be specified. Acquisition will be performed only for the specified satellites. An empty list (default) starts a search for all available satellites 1–32.

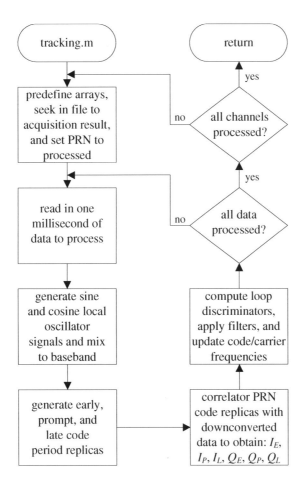

FIGURE A.3. GNSS tracking flow diagram.

acq_searchBand Specifies the frequency band in which to search for satellite signals, kHz integer. It is centered around the IF. While searching for available signals the acquisition function uses steps of 0.5 kHz.

acq_threshold Determines the threshold of the signal detector.

The output is an array structure acqResults containing search results for all satellites specified in acq_satelliteList. If a satellite signal is detected, the field signalDetected is set to 1 for that particular satellite.

A.4   Tracking Function

This function tracks the GPS signals allocated to each channel; see Figure A.3.

The function takes the following parameters: a block of the recorded signal from the front end, structure channel, sine, cosine, and C/A code tables. The function processes the block of the samples and returns two structures: tracking results trackResults and an updated structure channel.

Structure channel is used to pass initial information for each channel and is also used to store information on the current channel. The second purpose of this structure is to make the tracking continuous. In this way, processing of two or more signal blocks can be made continuous. The structure contains current (from last processed ms) information on the carrier frequency, code phase, PRN number of the track satellite, temporal values for the loop filters, and local signal generators.

The parameters for carrier tracking are contained in the settings structure

PLL_dampingRatio Damping ratio.

PLL_noiseBandwidth Noise bandwidth of the PLL.

The list below is the code tracking specific variables contained in the settings structure:

DLL_CACorrelatorSpacing Spacing between the early and late correlators, unit of chip.

DLL_dampingRatio Damping ratio for the delay lock loop.

DLL_noiseBandwidth Noise bandwidth of the delay lock loop.

The structure trackResults is the main output from tracking function. It contains results for all channels and for each ms of the processed block: information about signal properties (carrier frequency and code phase) and outputs from all six correlators and the loop discriminator.

The output from the tracking code is used as input for the postNavigation function. Additional information is used to plot tracking results and to analyze performance of the receiver. Execute command plotTracking to plot results for any individual channel.

A.5   Function postNavigation

The function starts by finding bit transitions and preamble locations. Then the bit values are obtained. The ephemerides are decoded. This involves only information from subframes 1, 2, and 3. Also decoding of subframes 4 and 5 may be included; see Figure A.4.

Next the function calls the pseudorange measurement function and computes position coordinates. Pseudorange and position computations are done covering a specified period described in the receiver settings.

The input for the function is the tracking results and settings structure and the output is pseudoranges and receiver coordinates.

The function postNavigation reads the following variables:

navSolPeriod Tells the software how often pseudoranges and position should be computed.

FIGURE A.4. Flow diagram for the position computation.

elevationMask Satellite elevation mask. Sets the minimum elevation angle for a satellite to be included in the computation of the position solution. Signals from satellites at low elevation angles are contaminated by large atmospheric errors.

UTMzone UTM Zone to be used for the coordinate transformation (ECEF to UTM). This is an integer that depends on the location of the receiver.

truePosition If an accurate position of the receiver antenna is known, then the **E**asting, **N**orthing, and **U**pping coordinates of the antenna can be specified. These coordinates will be subtracted from the ones computed by the software receiver, and the result will be plotted.

You may enter (E, N, U) either as approximate coordinates or as zeros.

A.5.1   Pseudorange Computation

This function computes pseudoranges for all tracked satellites. Pseudoranges are not computed for nontracking channels, for channels that have undetected preamble, or for channels that have detected a parity error in the navigation data; see Figure A.5.

The input for the function is the output of the in-phase prompt correlator. The function also needs precise code-phase information. Typically, both inputs are read from the structure trackResults.

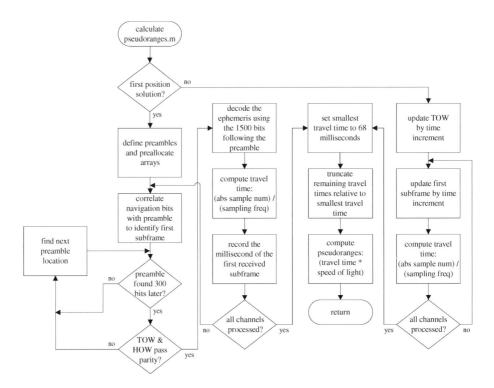

FIGURE A.5. Flow diagram for computing pseudoranges.

The output of the function is a set of pseudoranges and the signal transmission time (SOW) for all measured pseudoranges.

A.5.2 Position Computation

The function leastSquaresPos computes the position of the receiver from the measured pseudoranges. If for some reason the entire ephemeris for a satellite is not available, the pseudorange is excluded from the computation (see Figure A.6).

At the first computation, the elevation angles of all satellites are set to the maximum. This is necessary to estimate the receiver position and to compute the true elevation angles of all satellites. All subsequent computations of position will exclude pseudoranges from satellites with elevation angles lower than elevationMask.

The function postNavigation uses several other functions that are slightly modified versions of functions from the Easy Suite; see Borre (2003).

At the end of the postNavigation function, the ECEF coordinates are transformed to UTM and geodetic coordinate systems. The results are stored in structure navSolutions.

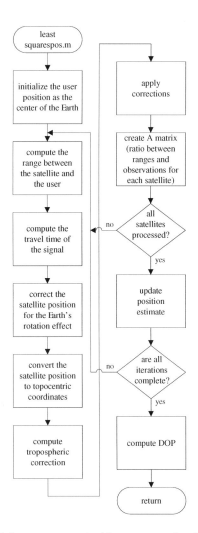

FIGURE A.6. Least-squares position computation flow diagram.

B
GNSS Signal Simulation

B.1   GPS Signal Simulation

When implementing the signal processing parts of the GPS receiver, it is necessary to have some data available for testing their functionality. The final goal is to have a GPS receiver working in real time on data obtained from a GPS antenna through an RF front end and an ADC. In the phase of developing the signal processing algorithms, however, it is not optimal to use real sampled data. The main reason for this is that it is impossible to control the properties of the received and sampled GPS signals. It is not only impossible to control the received signals, it is also impossible to know all the properties of the received signals.

The solution to this problem is to use simulated data. In lack of a suitable GPS signal simulator, it may be necessary to implement one.

A useful L1 GPS signal simulator should include the following global parameters associated with the downconversion and sampling of the signal:

Intermediate Frequency It should be possible to input the value of the IF. The IF will then be the reference frequency to which the Doppler shift of the satellite signals should be compared.

Sampling Frequency It should be possible to input the value of the sampling frequency used to sample the GPS signals.

With the possibility of setting these parameters, it would be possible to test the algorithms with simulated data with the same properties as the data sampled from a GPS antenna through an RF front end.

Below we list the properties of a GPS signal from one satellite—see Chapter 2 for additional details about the satellite signals:

PRN The pseudorandom noise number corresponding to the satellite. This number indicates which of the C/A codes should be used.

Doppler The Doppler count is the frequency deviation from the IF. The Doppler count is directly associated with line-of-sight dynamics between the satellite and the receiver.

Code Phase The code phase is the time alignment of the PRN code in the received data.

P(Y) Code In addition to the C/A code, the L1 signal contains the P(Y) code. This code is modulated onto the carrier wave as a quadrature component while the C/A code is present in the in-phase component.

Data Bits The navigation data bits are phase-modulated onto the carrier wave with a frequency of 50 Hz.

Signal Strength Due to the long signal path from the satellite to the receiver in combination with a low power transmitter at the satellite, the received signal is very weak. An additional property of the GPS signal is the signal-to-noise ratio (SNR), the ratio between the signal and the noise originating from the signal path.

The signal simulator should meet the above description of the global parameters and the GPS signal properties. For each satellite it should be possible to define PRN, Doppler, and code phase. The P(Y) code is not relevant to the acquisition and tracking algorithms. As a result of that, this component is simplified in the simulator. The P(Y) code is simulated as a digital signal alternating between -1 and 1 with the P(Y) code rate of 10.23 MHz. In most cases, the navigation data are not relevant to the acquisition and tracking algorithms. So by default the navigation data signal is also simulated as a digital signal alternating between -1 and 1 with the navigation data rate of 50 Hz. This gives a data bit transition each 20 ms. If a real navigation data sequence is needed it should be possible to provide it to the simulator.

B.2   Simulink Implementation

The GPS signal simulator could be implemented in any software development tool. The only demands to the program, except the possibility of implementing the needed functionalities, are

- It should be easy to change parameters of a simulation.

- It should be possible to save the simulated data in a file that can be read by MATLAB.

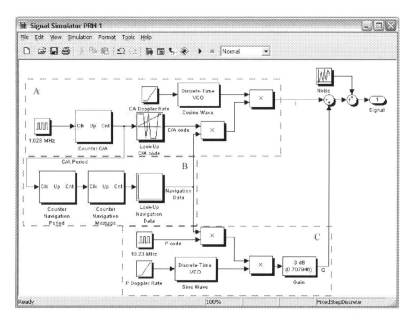

FIGURE B.1. The complete GPS signal simulator for one satellite implemented in Simulink. Part A generates the C/A code, Part B contains the navigation data generation, and Part C generates the simplified P(Y) code.

The MATLAB tool Simulink was chosen as the development tool for the GPS signal simulator. The main reason for this choice is that Simulink has a very intuitive user interface combined with its numerous features. One important feature in this case is that Simulink works perfectly together with ordinary M-files.

The Simulink implementation is a two-level design. The upper level is where each satellite contributing to the resulting signal is initialized. This is also where the data file in which the data should be saved is defined. The lower level is the implementation of each satellite. This is where the signal originating from each satellite is generated. The signals should include the following four components:

- C/A code
- Navigation data
- P(Y) code
- Noise.

Figure B.1 shows a screen shot of the lower level of the Simulink implementation representing the signal generation in a satellite.

The following will describe the four above-mentioned parts of the signal generator.

B.2.1   C/A Code Generation

The generation of the C/A code component of the GPS signal is marked with A in Figure B.1. Beginning from the left side of the figure, the source of the C/A code generator is the oscillator. This oscillator produces a squared pulse with a

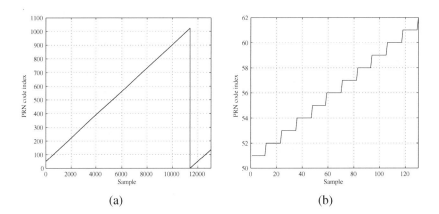

(a)   (b)

FIGURE B.2. C/A code counter output. (a) A complete cycle through all indexes of the code. (b) The first samples of the counter output. The initial value of the counter is the PRN code phase. In this case, the code phase is 51.

frequency of 1.023 MHz corresponding to the chipping rate of the C/A code. This pulse is used as input to a counter.

This counter is designed to count from 0 to 1022, referring to the 1023 chips in one PRN sequence. The counter increases its value each time a falling edge occurs in the pulse signal, that is, with a frequency of 1.023 MHz. The initial value of the counter is set by the PRN code phase, which is provided to the simulation as a parameter. The PRN code phase shifts the time alignment of the PRN code. Figure B.2 shows the output from the counter. These plots are made from a simulation with a sampling frequency of 12 MHz. A sampling frequency of 12 MHz causes 12,000 samples to last 1 ms, corresponding to one complete PRN code period.

Figure B.2a shows the output of the counter in a complete PRN code period. At sample zero the output of the counter corresponds to the PRN code phase. When the output reaches 1022 at approximately sample 11,000, the output is reset to its initial value 0. Figure B.2b is a close-up on the first samples of the output from the counter. The code phase is the initial value of the output, which in this case is 51. This figure shows the step function of the counter. Every time a chip period has passed the output, the counter increases by 1. When sampling with 12 MHz, a chip period of 1 ms/1023 lasts approximately 12 samples.

The output from the counter is provided as input to the next block of the C/A code generation part. This block is a so-called look-up table which is actually just an array indexed by the input to the block. In this case, the look-up table is a two-dimensional array containing the 32 possible PRN codes; see Section 2.3 for details on PRN codes. The first dimension indicates which of the 32 PRN codes should be used. This is the satellite supplied to the simulation. The second dimension indicates which of the 1023 chips that should be accessed. This is the input to the look-up table block. In this case, the array is input to the simulation from the MATLAB workspace. That is, before starting the simulation, the array with the PRN codes has to be generated. The implementation of the PRN code generator is described in Section 6.2. The resulting C/A code signal is shown in Figure B.3.

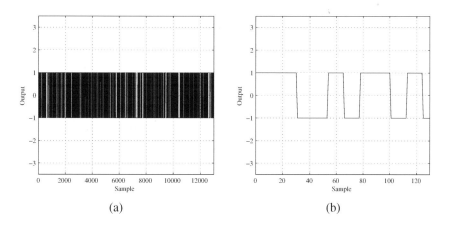

FIGURE B.3. C/A code. (a) A complete cycle through a complete PRN code period. (b) The first samples of the C/A code.

The sampled C/A code is multiplied with the sampled navigation data in the next block. The navigation data is also a binary data sequence consisting of -1's and 1's. The sampled navigation data generation will be described in the following section.

When the C/A code is combined with the navigation data, it must be modulated onto a carrier. This carrier is generated as a cosine wave in a voltage-controlled oscillator (VCO). The VCO has an input from a ramp block. This block simulates the change in frequency as a function of the Doppler rate. If the Doppler rate is zero, the frequency of the VCO remains constant.

The BPSK modulation of the C/A code and data signal is implemented as a multiplication of the signal with the generated carrier wave.

B.2.2   Navigation Data Generation

The generation of the navigation data component of the GPS signal is indicated within part B in Figure B.1. As mentioned earlier, the bit rate of the binary navigation data sequence is 50 Hz. In the simulator, the generator signal to the navigation data sequence is obtained from the output of the counter in the C/A part. As seen in Figure B.2a, the counter resets after running through a complete PRN code sequence. As mentioned, this period is 1 ms corresponding to a rate of 1 kHz. To obtain a 50 Hz signal from the signal supplied by the C/A counter, the first block in the navigation data part is a counter, which increases its value on every falling edge of the input signal. The counter resets when it reaches 20, that is, every 20 ms corresponding to 50 Hz. That is, the output from the first counter in the navigation data part of the simulator supplies a 50 Hz signal to the next block.

The next block is also a counter. This counter works in the same way as the counter in the C/A part. That is, it increases its output value with every falling edge of the input signal, and it resets at the end of a period. In this case, the period is set to be one navigation data frame corresponding to 1500 bits.

The next block is a look-up table much similar to that of the C/A part of the simulator. In this case the look-up table contains a two-dimensional array, where the first dimension selects whether or not there should be any navigation data transitions in the data. The second dimension is the index of the navigation bit to be the output from the look-up table. In case no navigation bit transitions are wanted, the look-up table always outputs the value 1. In the other case where navigation bit transitions are wanted, the look-up table outputs alternating −1's and 1's.

The output from the navigation data part is provided to the C/A and the P code part through a multiplicator.

B.2.3   P Code Generation

The generation of the P code component of the GPS signal is indicated within Part C in Figure B.1. As mentioned before, the P code component is not relevant to the acquisition and tracking algorithms and should thus only be simulated as a squared pulse alternating between −1 and 1. In the GPS signal simulator this signal is implemented as a squared pulse generator with a frequency corresponding to the chipping rate of the P code, which is 10.23 MHz.

In the second block of the P code part, this signal is combined with the navigation data sequence. This is done in a similar way as the C/A part by multiplication of the two signals.

The carrier generation part in the lower left corner is also almost similar to the carrier generation part in the C/A part. The only difference here is that the VCO has a 90° phase shift compared to the other VCO. The result of this is that the P code VCO generates a sine wave compared to the C/A VCO's cosine wave. The combined P code and navigation data signal is BPSK modulated onto the carrier in the multiplication block.

The final block of the P code part is a gain block. This block decreases the modulated P code signal by 3 dB in the same way as it is done in the signal generator in the GPS satellites.

B.2.4   Combining the Signal Components

The last part of the GPS signal simulator is located at the right-most part of Figure B.1. Here, the two modulated codes are combined, resulting in a complete GPS signal. The two components are simply added together as in-phase and quadrature components of the final signal.

The last part of the signal simulator is the addition of noise. This noise generator block is indicated with green in Figure B.1. The amount of noise is selected from an input to the simulation. The resulting noise from the noise generator is added to the GPS signal.

B.2.5   Upper-Level Implementation

When the lower level of the Simulink implementation of the GPS signal generator has been implemented, the upper-level implementation can be designed. The

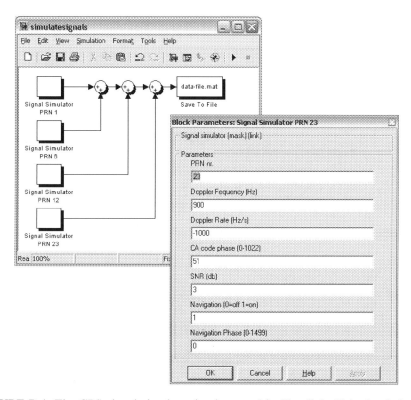

FIGURE B.4. The GPS signal simulator implemented in Simulink. This simulation includes four different satellites and saves the data in a file. The other window pops up when one of the signal simulator boxes are double-clicked. This window is used to set all values for one of the satellites in the simulation.

upper-level implements the different satellites needed in the simulation and selects the file in which the simulated data should be saved. Figure B.4 shows an example of how the upper level of the signal simulator could be implemented. In this case, the simulator contains four different satellites. Each of the satellites has different values of PRN, PRN code phase, Doppler shift, and Doppler rate that are typed into a window that pops up when one of the boxes in Part A is double-clicked.

B.3   Galileo Signal Generator

In order to further study the Galileo signals, we have implemented a simple signal simulator in Simulink.

Figure B.5 shows the Simulink model of the Galileo signal simulator. The present version is made of standard Simulink blocks, and it does not have any custom blocks (called subsystems in Simulink). The gray blocks in the model are part of the generator, and the white blocks are used to visualize the generated signal. The white blocks do not contribute in any way to the signal generation. The

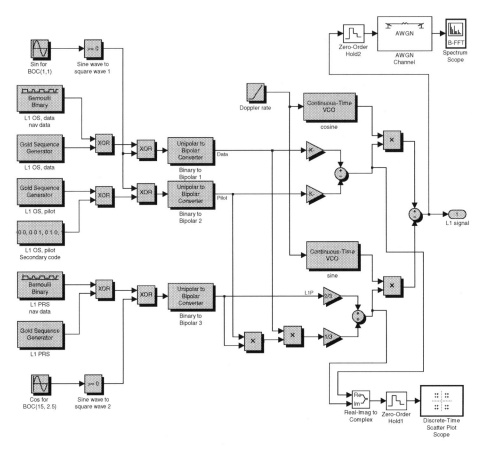

FIGURE B.5. The Simulink model that generates the L1 signal. The gray blocks generate the Galileo signal, and the white blocks are used to visualize the signal.

generator directly implements the BOC, CASM, and other steps described earlier in this chapter. It is easy to see CASM constants and math operations in the model that correspond to (3.2)–(3.4).

The model can be divided into three parts from left to right (excluding the white blocks). The first part consists of the generators of PRN, BOC signal, and random bits of navigation messages. This is the left part of the model. It makes all three navigation signals at the baseband (from the top to the bottom): data channel, pilot channel, and restricted access channel.

The center part of the diagram converts binary signals of the left side to bipolar ±1 signals. Simulink PRN generators are generating binary data and the bipolar signals are required for the CASM algorithm.

The CASM part is implemented on the right side of the model. The output of the model is Galileo L1 signal at the L1 frequency. The zero-order hold blocks are used to sample continuous signal. The sampling frequency is 110.5 MHz, and the result of sampling the IF is about 28 MHz. The signal spectrum after sampling is shown in Figure B.6. It clearly shows the two main lobes of BOC(1,1) at the center of the spectrum. The two OS signals are sharing this part of the spectrum. The two

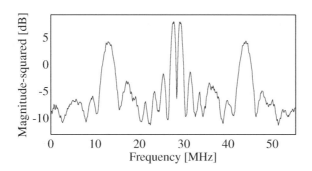

FIGURE B.6. Spectrum for generated Galileo L1 signal.

wide lobes at the sides of the spectrum originate from the restricted access signal BOC(15,2.5).

Our model does not save the signal from the generator. The output of the generator must be sampled and directed to a file or a variable in MATLAB workspace. The technique is similar to the implementation of the FFT plot visible at the top right corner of the diagram. A zero-order hold block is needed to sample the signal. Then one of the signal sinks must be inserted from the Simulink block library depending on whether it has to be a file or a variable. The Additive White Gaussian Noise (AWGN) block is used to add noise to the signal.

The current model only generates the signal for one satellite at a time. It can be used as a subsystem in Simulink. So several such generators must be connected to generate signals of a constellation of satellites. Section B.2.5 describes how that can be implemented.

The current model is very basic and must be updated comprehensively in order to account for properties and propagation environment of the real signals. First a more complicated signal propagation model must be added which accounts for signal amplification at the satellite, signal losses in free space, atmosphere, antennas, and the front end. Next the model must be modified to include Doppler effects. The rate of all generators (data, PRN, subcarriers, and the carrier) must be controlled simultaneously and all intersignal phases must be preserved. This requires custom implementations of these blocks. An excellent example of an advanced generator is explained in the Mathworks webinar "Implementing a GPS Receiver with DSP and FPGA Hardware Using Simulink and Related Tools."

B.4   Differences in Processing GPS and Galileo Signals

Originally this text was planned to deal with both GPS and Galileo on equal terms. It soon became clear that this ideal goal had to be modified. It turned out to be difficult to get reliable detailed information about Galileo signals; in addition, the necessary testing of the software implementation also became difficult to carry out.

We therefore decided to list the differences between GPS and Galileo signals and next mention how to deal with them in a future software implementation.

B.4.1   Signal Differences

This listing only considers Galileo L1 OS and GPS C/A on L1. First the signal differences are analyzed. The following paragraphs outline the differences in the receiver signal processing.

Signal Types GPS has one public signal and one encrypted, restricted access signal. Galileo will have three signals: two public signals, called Open Service (OS) signals, and one encrypted, called a Public Regulated Service (PRS) signal. Only the OS signals will be considered in the following. One of the OS signals, the data channel, will contain navigation data (ephemerides, almanac, and additional information). The pilot channel will not be modulated with navigation data. It will only be modulated with a short sequence of bits (a secondary code with a code length of 25 chips), which will be repeated all times.

Spreading Codes GPS uses a spreading code with 1023 chips, whereas Galileo will use spreading codes with lengths of 4096 chips. The chipping rate is the same for GPS and the Galileo OS signals, 1.023 MHz, but all Galileo codes on L1 are combined with a subcarrier signal (BOC signals). The subcarrier rate is 1.023 MHz for both the L1 OS data and pilot signals. It is expected that a similar type of signal will be used in the modernized GPS (most likely by the GPS III program). The transmitted GPS L1 signals are bandwidth limited to 20×1.023 MHz whereas the Galileo L1 signals are limited to 40×1.023 MHz.

The Galileo PRN codes are not yet published. Depending on the final choice of codes, two techniques exist to generate the PRN codes. One is to use linear feedback shift register generators similar to those used in GPS, but with longer registers. The second is to use memory codes that may be pregenerated and stored in memory.

Recent Galileo signal developments indicate that additional codes may be used on top of BOC coding to improve signal properties and signal tracking performance; see Hein et al. (2005).

Data Modulation The data modulation process is the same in GPS and Galileo. The BOC signal is multiplied by the data signal (the XOR operation is the equivalent for binary logic signals). The pilot signal is using the same technique to modulate the BOC signal by the secondary code. The navigation data rate on the data channel is 250 Hz. It is likely that the data rate of the secondary code on the pilot channel will be similar to this.

Data Structure The Galileo system will use a superframe, frame, subframe construction similar to GPS. Subframes will have a unique word facilitate synchronization to the start of the subframe (similar to the preamble in GPS). The unique synchronization word is followed by the data part, a checksum field, and tail bits. It is expected that the construction of the data part will be different from that of the GPS messages.

Ephemerides and Almanac The satellite orbit parameters have the same field size and scale in both systems. The time parameters have different field size and scales (except clock correction coefficients in the almanac).

Synchronization Word (Preamble) GPS is using an 8-bit (symbol) pattern. Galileo is likely to use a 10-symbol pattern.

Error Detection Galileo will use cyclic redundancy check (CRC) to detect data corruptions inside subframes. It is expected that the CRC will be computed over the 24 data bits in the subframe. The specification does not indicate that any bits from the previous subframe should be used in computation of the CRC.

Channel Coding In addition to CRC, Galileo will use forward error correction (FEC) to detect data corruptions and correct corruptions to a certain extent. This will facilitate correction of a much larger amount of corruptions compared to GPS where only one bit per subframe can be corrected. Block interleaving will be used to make the Galileo data even more corruption resistant.

FEC is used already in WAAS and EGNOS signals.

Data Authentication Galileo is likely to use a data authentication technique to make the GNSS signal tracking secure. The purpose of the data authentication is to provide means for the user to distinguish genuine Galileo signals from simulated signals (the technical term is signal spoofing). Signal spoofing is an intentional malicious provision of faulty signals that can lead to malicious location spoofing.

Only certified receivers will be able to decode authentication information, but the authentication will be provided by the OS signals. P(Y) code encryption is the means of signal authentication in GPS.

Modulation Galileo uses BOC(1,1) modulation (which in effect perform manchester encoding of the data and pilot channel) and the CASM multiplexing scheme to combine three signals into a hexaphase representation whereas GPS uses BPSK.

Time Reference Galileo will use a reference time called Galileo System Time (GST) whereas GPS uses GPS Time (GPST). GPS Time is a composite clock; the average of a number of GPS clocks computed in a Kalman filter. Galileo System Time is a master clock; the output of a steered active H-maser. Galileo System Time will be steered to International Atomic Time (TAI) at the Galileo Precise Time Facility (PTF). GPS Time is steered to a real-time representation of Coordinated Universal Time (UTC) by the

U.S. Naval Observatory (USNO). The offset between TAI and UTC is an integer number of seconds.

Satellite Constellation The GPS baseline system is specified for 24 satellites; however, the system currently consists of more than 24 satellites. The constellation contains 6 orbital planes inclined 55° to the equator. Each plane contains 4–5 active satellites. The satellite orbit altitude is 20,183 km from mean surface of the Earth and the satellites have an orbital period of 11 hours 58 minutes.

The Galileo space segment will comprise 30 satellites in a Walker constellation with 3 orbital planes inclined 56° to the equator. Each plane will contain 9 operational satellites (for a total of 27 active satellites) equally spaced 40° apart plus 1 inactive spare. The satellite orbit altitude is 23,222 km. This corresponds to a constellation repeat cycle of 10 days during which each satellite has completed 17 revolutions.

B.5   Differences in Signal Processing

The following is an outline of the receiver signal processing differences.

Signal at IF The Galileo OS signals require two times wider signal bandwidth for the main lobes, i.e., 2×2.046 MHz, compared to 1×2.046 MHz for GPS. In practical applications it might be prudent to use at least twice this minimum bandwidth to ensure less distortion of the ACF of the received signals. Thus, a GPS front end with narrow bandwidth might need an update to be ideal for Galileo reception.

If modifying a GPS front end to increase reception bandwidth, care should be taken not to let mixer spectrum mirrors overlap the original signal spectrum. Thus, the IF frequency might need adjustments. This in turn means that the ranges of possible sampling frequencies might be affected since care should be taken to avoid aliasing occuring at $f_{alias} = f_{sampling} - f_{signal}$.

Otherwise there should be no need for changes at the radio/IF receiver part since both GPS and Galileo signals share the L1 carrier frequency and are transmitted at comparable signal power levels. Front ends for Galileo signals will not need any modifications to receive GPS signals.

Differences Common to Acquisition and Tracking The main difference between GPS and Galileo, which is common to acquisition and tracking, is the different spreading codes. It is hard to outline the precise differences, because the Galileo PRN generator details are not published yet. But in any case there will have to be two different PRN generators for GPS and Galileo. For Galileo the output of the PRN generators must be modulated onto a square wave to make the BOC signal. The Galileo codes are 4 times longer than the GPS codes. Since the chipping rate is the same, this specifies a

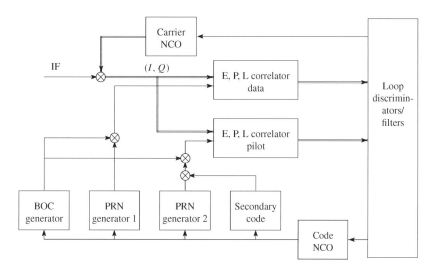

FIGURE B.7. Generic L1 OS tracking architecture.

minimum correlation time (dwell time) of 4 ms for a complete code period of correlation for both acquisition and tracking.

The Galileo acquisition and tracking might need additional techniques to acquire/track the right peak of the ACF if the L1 OS signals will use the proposed CBCS coding scheme, confer Hein et al. (2005). Depending on the configuration of the CBCS coding scheme, ACFs with multiple, local extreme might occur.

A relatively simple false lock detection consisting of checking the ACF signal power in the vicinity of the tracked point could be implemented. This could be done with either extra correlator arms, situated "very early" and "very late" with respect to the early, prompt, and late arms, or it could be done by periodically jumping to neighboring code phases with the existing early, prompt, and late correlators.

Acquisition of Galileo Signals The acquisition of Galileo signals can be carried out as with GPS signals except from the differences mentioned in the previous subsection. In the case of the multiple peak correlation function, the local extremes could be ignored in the acquisition process due to the limited code phase resolution. The check for false locks should then be carried out when the handover from acquisition to tracking has taken place and the DLL has converged.

The dwell time dictates the maximum step for the frequency search. For the minimum length of 4 ms the maximum frequency step is 250 Hz (was 1 kHz for the 1 ms dwell time). For an improved performance a 125 Hz would be a better choice.

Tracking The high level architecture of the tracking block is almost exactly the same in the basic BOC(1,1) case as in GPS. The main differences are the new code generators (BOC, PRN, and secondary code) and longer minimum dwell times. A generic L1 OS tracking architecture is shown in Figure B.7.

The downconversion from IF to baseband is done with the usual carrier NCO and mixer. A separate PRN code generator (or look-up table depending on the final implementation) is used for the local data and pilot channel PRN code replicas. Both local codes are combined with the BOC subcarrier. The pilot channel code is combined with the slow varying secondary code. All code generators are controlled by the strobe signal from the code NCO, although the secondary code generator uses a divided version of this timing signal. Suitable discriminators and loop filters provide the control signals for both NCOs as usual.

Both the data and pilot signals must be tracked together (combined tracking) to maximize receiver performance.

Data Demodulation and Decoding The demodulation will need an updated preamble search function. The code will have to employ a deinterleaver to put the received symbols (FEC encoded bits) in the right order. The MATLAB communication toolbox has built-in functions for interleaving and deinterleaving.

A Viterbi decoder must be used to decode the symbol stream encoded by the FEC algorithm. The MATLAB communication toolbox has a built-in function vitdec to decode the FEC stream. The subframes have 6 tail bits set to zero to "initialize" the Viterbi decoder at the start of the next subframe. Therefore, the decoding can commence at the start of any subframe without prior knowledge of the navigation data stream.

The CRC check must be computed to detect any errors. The algorithm is very similar but not identical with that of GPS.

The actual locations of all transmitted data parameters in the data messages will be different. Also, some data will be transmitted in the Galileo system that is not transmitted in GPS. Therefore, new code must be written to decode navigation messages.

Decoding of ephemerides and almanac data is very similar to that of GPS once the data fields are read from the data messages. Decoding of time-related fields will need minor changes due to new width of the data fields and scale factors.

Position Computation The position computation is exactly the same as in GPS once the time-related data and ephemerides are decoded. There is a small difference in the coordinate systems (few centimeters) between GPS and Galileo. Care must be taken to handle this properly in case of very precise measurements.

Bibliography

Akos, Dennis (1997). *A Software Radio Approach to Global Navigation Satellite System Receiver Design.* Ohio University, Athens, OH.

Akos, D. M., Stockmaster, M., Tsui, J. B. Y. & Caschera, J. (1999). Direct bandpass sampling of multiple distinct RF signals. *IEEE Transactions on Communications*, 47(7):983–988.

Anonymous (1997). *World Geodetic System 1984, Its Definition and Relationships with Local Geodetic Systems.* National Imagery and Mapping Agency, 3rd edition, St. Louis, MO.

Anonymous (2000). Application note: Selecting an A/D converter. Texas Instrument, focus.ti.com/lit/an/sbaa004/sbaa004.pdf.

Anonymous (2005). L1 band part of Galileo Signal in Space ICD (SIS ICD). Galileo Joint Undertaking, http://www.galileoju.com/page.cfm?voce= s2&idvoce=64&plugIn=1.

Balanis, Constantine A. (1996). *Antenna Theory: Analysis and Design.* John Wiley & Sons, Inc., 2nd edition, New York, NY.

Bastide, F., Akos, D., Macabiau, C. & Roturier, B. (2003). Automatic gain control (AGC) as an interference assessment tool. In *16th International Technical Meeting of the Satellite Division of the Institute of Navigation*, pages 2042–2053, Portland, OR.

Best, Roland E. (2003). *Phase-Locked Loops: Design, Simulation, and Applications*. McGraw-Hill, 5th edition, New York, NY.

Betz, John W. (2002). Binary offset carrier modulations for radionavigation. *Navigation*, 48:227–246.

Borre, Kai (2003). The GPS easy suite—MATLAB code for the GPS newcomer. *GPS Solutions*, 7:47–51.

Chung, B.-Y., Chien, C., Samueli, H. & Jain, R. (1993). Performance analysis of an all-digital BPSK direct-sequence spread-spectrum IF receiver architecture. *IEEE Journal on Selected Areas in Communications*, 11(7):1096–1107.

Dixon, R. C. (1984). *Spread Spectrum Systems*. John Wiley & Sons, 2nd edition, New York, NY.

Forssell, Börje (1991). *Radionavigation Systems*. Prentice-Hall, Englewood Cliffs, NJ.

Gold, Robert (1967). Optimal binary sequences for spread spectrum multiplexing. *IEEE Transactions on Information Theory*, 13(4):619–621.

Golomb, S. (1982). *Shift Register Sequences*. Aegean Park Press, Laguna Hills, CA.

Gromov, K., Akos, D., Pullen, S., Enge, P. & Parkinson, B. (2000). GIDL: Generalized interference detection and localization system. In *13th International Technical Meeting of the Satellite Division of the Institute of Navigation*, pages 447–457, Salt Lake City, UT.

Haykin, S. (2000). *Communication Systems*. John Wiley & Sons, 4th edition, New York, NY.

Hein, Guenter W., Avila-Rodriguez, Jose-Angel, Ries, Lionel, Lestarquit, Laurent, Issler, Jean-Luc, Godet, Jeremie & Pratt, Tony (2005). A candidate for the Galileo L1 OS optimized signal. In *18th International Technical Meeting of the Satellite Division of the Institute of Navigation*, pages 833–845, Long Beach, CA.

ICD-GPS-200 (1991). Interface control document. ICD-GPS-200, Arinc Research Corporation, 11 770 Warner Ave., Suite 210, Fountain Valley, CA.

ICD-GPS-705 (2002). Interface control document: Navstar GPS space segment/navigation L5 user interfaces. US DOD.

Kaplan, Elliott D. & Hegarty, Christopher J., editors (2006). *Understanding GPS, Principles and Applications*. Artech House, 2nd edition, Boston, MA.

Martin, Nicolas, Leblond, Valéry, Guillotel, Gilles & Heiries, Vincent (2003). BOC(x,y) signal acquisition techniques and performances. In *Proceedings of ION GPS/GNSS 2003*, pages 188–197, Portland, OR.

Mattos, Philip (2004). Acquiring sensitivity to bring new signal indoor. *GPS World*, May:28–33.

Nunes, Fernando D., Sousa, Fernando M. G. & Leitão, José M. N. (2004). Multipath mitigation technique for BOC signals using gating functions. In *2nd ESA Workshop on Satellite Navigation User Equipment Technologies, NAVITEC '2004, 8–10 December*, ESTEC, Noordwijk.

Oppenheim, A. & Schäfer, R. (1999). *Discrete-Time Signal Processing*. Prentice-Hall, Englewood Cliffs, NJ.

Parkinson, Bradford W. & Spilker Jr., James J., editors (1996). *Global Positioning System: Theory and Applications*, volume 163 of *Progress in Astronautics and Aeronautics*. American Institute of Aeronautics and Astronautics, Inc., Washington, DC.

Shanmugan, K. Sam & Breipohl, A. M. (1988). *Random Signals: Detection, Estimation and Data Analysis*. John Wiley & Sons, New York, NY.

SPS (1995). Global positioning system standard positioning service signal specification. U. S. Department of Defense.

Strang, Gilbert & Borre, Kai (1997). *Linear Algebra, Geodesy, and GPS*. Wellesley-Cambridge Press, Wellesley, MA.

Straw, R. Dean, editor (2003). *The ARRL Antenna Book: The Ultimate Reference for Amateur Radio Antennas, Transmission Lines and Propagation*. American Radio Relay League, 20th Bk & Cr edition, Newington, CT.

Tsui, J. (2000). *Fundamentals of Global Positioning System Receivers: A Software Approach*. John Wiley & Sons, New York, NY.

Winkel, Jón (2005). BCS spectrum calculation. To be published.

Winkel, Jón Ólafur (2000). *Modelling and Simulating GNSS Signal Structures and Receivers*. Universität der Bundeswehr München, Neubiberg.

Ziemer, Rodger E. & Peterson, Roger L. (1985). *Digital Communications and Spread Spectrum Systems*. MacMillan, New York, NY.

Index

Printed in the United States
By Bookmasters